EYE AND VISION RESEARCH DEVELOPMENTS SERIES

UNDERSTANDING CORNEAL BIOMECHANICS THROUGH EXPERIMENTAL ASSESSMENT AND NUMERICAL SIMULATION

EYE AND VISION RESEARCH DEVELOPMENTS SERIES

Eye Cancer Research Progress
Edwin B. Bospene (Editor)
2008. ISBN: 978-1-60456-045-9

Non-Age Related Macular Degeneration
Enzo B. Mercier
2008. ISBN: 978-1-60456-305-4

Optic Nerve Disease Research Perspectives
Benjamin D. Lewis and Charlie James Davies (Editors)
2008. ISBN: 978-1-60456-490-7

Optic Nerve Disease Research Perspectives
Benjamin D. Lewis and Charlie James Davies (Editors)
2008. ISBN: 978-1-60741-938-9 (Online book)

New Topics in Eye Research
Lauri Korhonen and Elias Laine (Editors)
2009. ISBN: 978-1-60456-510-2

Eye Infections, Blindness and Myopia
Jeffrey Higgins and Dominique Truax (Editors)
2009. ISBN: 978-1-60692-630-7

Retinal Degeneration: Causes, Diagnosis and Treatment
Robert B. Catlin (Editor)
2009. ISBN: 978-1-60741-007-2

Retinal Degeneration: Causes, Diagnosis and Treatment
Robert B. Catlin (Editor)
2009. ISBN: 978-1-60876-442-6 (Online book)

**Eye Research Developments:
Glaucoma, Corneal Transplantation, and Bacterial Eye Infections**
Alan N. Westerhouse (Editor)
2009.ISBN: 978-1-60741-1772

Binocular Vision: Development, Depth Perception and Disorders
Jacques McCoun and Lucien Reeves (Editors)
2010. ISBN: 978-1-60876-547-8

Retinitis Pigmentosa: Causes, Diagnosis and Treatment
Michaël Baert and Cédric Peeters (Editors)
2010. ISBN: 978-1-60876-884-4

Color: Ontological Status and Epistemic Role
Anna Storozhuk
2010. ISBN: 978-1-61668-201-9

Color: Ontological Status and Epistemic Role
Anna Storozhuk
2010. ISBN: 978-1-61668-608-6 (Online book)

Coherent Effects in Primary Visual Perception
V.D. Svet and A.M. Khazen
2010. ISBN: 978-1-61668-143-2

Coherent Effects in Primary Visual Perception
V.D. Svet and A.M. Khazen
2010. ISBN: ISBN: 978-1-61668-496-9 (Online book)

Conjunctivitis: Symptoms, Treatment and Prevention
Anna R. Sallinger
2010. ISBN: 978-1-61668-321-4

Conjunctivitis: Symptoms, Treatment and Prevention
Anna R. Sallinger
2010. ISBN: 978-1-61668-443-3 (Online book)

Understanding Corneal Biomechanics through Experimental Assessment and Numerical Simulation
Ahmed Elsheikh
2010. ISBN: 978-1-60876-694-9

EYE AND VISION RESEARCH DEVELOPMENTS SERIES

UNDERSTANDING CORNEAL BIOMECHANICS THROUGH EXPERIMENTAL ASSESSMENT AND NUMERICAL SIMULATION

AHMED ELSHEIKH

Nova Science Publishers, Inc.
New York

Copyright © 2010 by Nova Science Publishers, Inc.

All rights reserved. No part of this book may be reproduced, stored in a retrieval system or transmitted in any form or by any means: electronic, electrostatic, magnetic, tape, mechanical photocopying, recording or otherwise without the written permission of the Publisher.

For permission to use material from this book please contact us:
Telephone 631-231-7269; Fax 631-231-8175
Web Site: http://www.novapublishers.com

NOTICE TO THE READER

The Publisher has taken reasonable care in the preparation of this book, but makes no expressed or implied warranty of any kind and assumes no responsibility for any errors or omissions. No liability is assumed for incidental or consequential damages in connection with or arising out of information contained in this book. The Publisher shall not be liable for any special, consequential, or exemplary damages resulting, in whole or in part, from the readers' use of, or reliance upon, this material. Any parts of this book based on government reports are so indicated and copyright is claimed for those parts to the extent applicable to compilations of such works.

Independent verification should be sought for any data, advice or recommendations contained in this book. In addition, no responsibility is assumed by the publisher for any injury and/or damage to persons or property arising from any methods, products, instructions, ideas or otherwise contained in this publication.

This publication is designed to provide accurate and authoritative information with regard to the subject matter covered herein. It is sold with the clear understanding that the Publisher is not engaged in rendering legal or any other professional services. If legal or any other expert assistance is required, the services of a competent person should be sought. FROM A DECLARATION OF PARTICIPANTS JOINTLY ADOPTED BY A COMMITTEE OF THE AMERICAN BAR ASSOCIATION AND A COMMITTEE OF PUBLISHERS.

LIBRARY OF CONGRESS CATALOGING-IN-PUBLICATION DATA

Elsheikh, Ahmed, 1943-
  Understanding corneal biomechanics through experimental assessment and numerical simulation / Ahmed Elsheikh.
     p. ; cm.
  Includes bibliographical references and index.
  ISBN 978-1-60876-694-9 (softcover)
  1. Cornea. 2. Human mechanics. I. Title.
  [DNLM: 1. Cornea--physiology. 2. Biomechanics. 3. Models, Biological. WW 220 E49u 2009]
  QP477.8.E47 2009
  617.7'19--dc22
                              2009044370

Published by Nova Science Publishers, Inc. ✦ *New York*

CONTENTS

Preface		ix
Chapter 1	Introduction	1
Chapter 2	Corneal Microstructure	3
Chapter 3	Corneal Topography	7
Chapter 4	Corneal Biomechanical Behaviour	9
Chapter 5	Numerical Modelling of Ocular Biomechanics	37
Chapter 6	Conclusion	51
References		53
Index		61

PREFACE

The Ocular Biomechanics Group was established in 2002 with one clear target; to develop a virtual reality model of the human eye that can be used effectively and reliably to predict ocular response to surgery, injury and disease. This ambitious, and seemingly illusive, target helped plan our activities over the last 6 years and will still be focusing our efforts as we strive to create the necessary knowledge using experimental methods, build the predictive tools using programming and analysis means, and validate the findings in both the laboratory and the clinic. This chapter presents an overview of our biomechanical studies from laboratory material characterisation to finite element numerical simulation. The chapter describes what has been achieved and points at the remaining gaps in our knowledge. It explains that while much remains unknown in ocular behaviour, we are now in a good position to use available knowledge to progress predictive modelling and use it in actual applications such as improving the accuracy of tonometry techniques, planning of refractive surgeries and design of contact lenses. The discussion focuses on the cornea, although scleral biomechanics receive some mention. The chapter also refers to microstructural, biomechanical and topographic studies conducted by other research groups. Coverage of these studies has been necessary to provide a more complete image of current understanding of corneal biomechanics.

Chapter 1

INTRODUCTION

The transparent cornea is a most important component of the outer ocular tunic. It provides a tough protective envelope for the ocular contents and helps give the eye its general shape, Figure 1. The anterior corneal surface accounts for about two-thirds of the optical power of the eye [1].

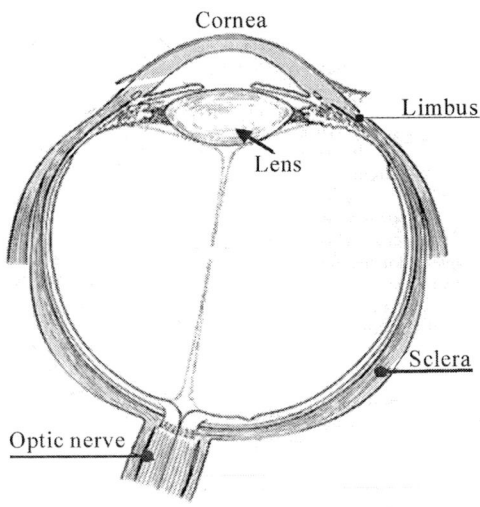

Figure 1. Cross-section of eye globe showing some of the main components [4].

This important role can be defined in terms of corneal shape, regularity and clarity, and is a function of its refractive index. The cornea's contribution to ocular image formation can be degraded by anomalies in shape produced by disease, injury or surgery. The ability to understand and predict corneal response to such conditions is therefore of great clinical importance [2, 3].

The past three decades saw a rapid growth in the use of numerical modelling to improve understanding of corneal biomechanics and to predict ocular response to the conditions met in tonometry and refractive surgery. Numerical models have improved in accuracy and in their representation of the natural conditions of the eye. In the 1980s and 1990s, it was common to assume linear elastic, and homogenous material behaviour and spherical topography in corneal numerical models [5-8]. Since then, and with the better understanding of corneal behaviour and topography achieved experimentally and clinically, these assumptions have become unnecessary [9]. The increase in computing power over these three decades also made it possible to produce and execute highly nonlinear corneal models with fine finite element meshes and with consideration of other ocular components [10, 11]. These highly expensive models, which required specialised mainframe computers in the past, can now be conducted on commonly available personal computers.

However, in spite of this progress, corneal numerical simulation had only a limited role in improving clinical practice. An example of this situation can be seen in the attempts to create numerical models of refractive surgery, where much attention has been given to simulate the procedures [12-23], but as yet no numerical model has been adopted in clinical practice to plan and optimise refractive surgeries. Offering a solution here might not be straightforward since part of the problem is related to the clinicians' reluctance to adopt tools and findings that come through a numerical route. However, much progress can be made by reviewing current modelling approaches, and concentrating on the areas where improvement is needed.

Recent developments in corneal microstructure detailing, material characterisation and topography mapping are offering valuable information for numerical modelling. Microstructural studies have identified the preferential orientation of stromal collagen fibrils [24-27] and the resulting anisotropy in corneal biomechanical behaviour [28, 29]. The corneal nonlinear stress-strain behaviour [14, 30, 31], viscoelasticity [32, 33], hysteresis [34], long-term creep [35] and stress relaxation [36] have been measured experimentally. The epithelium's contribution to corneal biomechanics has been quantified in relation to the stroma [37, 38]. More recently, the inter-lamellar adhesion within the stroma, under both shear [39] and lateral tension [40-42], has been assessed. Much work has also been done to determine the anterior and posterior asphericity [43-45], and the thickness distribution of the cornea [46, 48]. These properties offer new opportunities to make numerical modelling more representative of corneal behaviour, and potentially more reliable. But the question, which this chapter attempts to answer, is how best to use this knowledge, and whether it is essential for a numerical model to consider all these properties regardless of the application at hand.

The chapter starts with an overview of recent microstructural, topographical and experimental advances that provided significant improvements in our understanding of corneal biomechanics. A discussion is then presented on how to embed the experimentally-obtained material behaviour in numerical simulation. Within this discussion, a general introduction to finite element modelling and the simplifications involved in model construction is presented, followed by an assessment of the effects of simplifications on model output and whether these effects are acceptable. The chapter ends with concluding statements on the way forward for corneal numerical simulation.

Chapter 2

CORNEAL MICROSTRUCTURE

The human cornea consists of three distinct layers: an anterior epithelium, a posterior endothelium and a central stroma, Figure 2. Between these layers are specialised extracellular structures called Bowman's layer and Descemet's membrane. While the epithelium and endothelium are cellular, composed mainly of keratinocytes, the stroma has a distinctive composite structure with collagen fibrils embedded in an extracellular matrix rich in proteoglycans, glycoproteins and keratocytes.

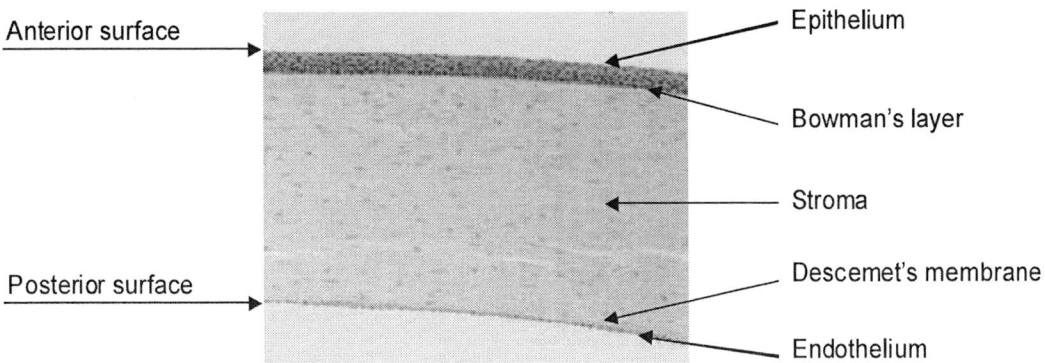

Figure 2. Cross-sectional view of the human cornea's layered construction.

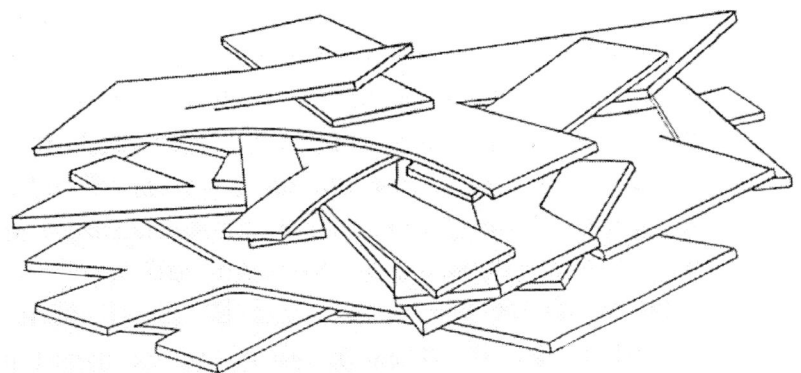

Figure 3. Interweaving of stromal lamellae.

The mechanical behaviour of the human cornea is dominated by the stroma that forms about 90% of its thickness [24, 49]. The stroma has a highly complex structure with 200-400 superimposed lamellae, set mostly parallel to the corneal surface, but increasingly interwoven towards the anterior surface [25, 50-52], Figure 3. The increased lamella interweaving possibly leads to variation in corneal mechanical properties and swelling resistance [53].

Stromal lamellae are composed of fine collagen fibrils regularly spaced and embedded in an extracellular matrix. The fibrils within each lamella run parallel to each other, but make large angles with neighbouring lamellae, Figure 4. Earlier research identified two preferential directions within which most fibrils run in the central human cornea, namely the superior-inferior and the temporal-nasal directions, possibly related to the action of the extraocular muscles [26, 27, 54-57]. Daxer and Fratzl estimated that 49% of fibrils are oriented in these two directions and 66% of the fibrils lie within the 45° sectors surrounding them [58]. Since the fibrils are thought to be the main load carrying elements of the stroma, their preferential orientation is likely to lead to anisotropic mechanical behaviour of the human cornea with the vertical and horizontal directions exhibiting superior stiffness (resistance to deformation) compared with other directions.

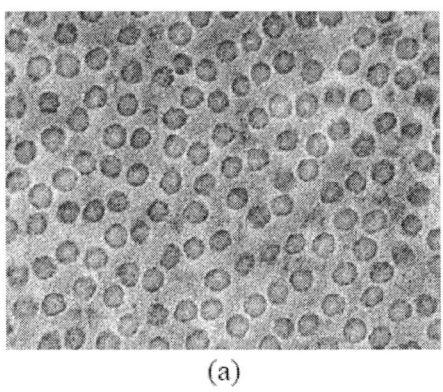 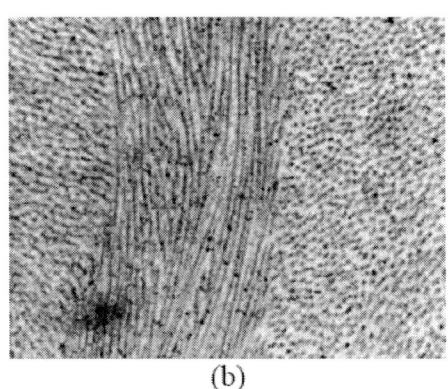

(a) (b)

Figure 4. TEM images of (a) collagen fibril distribution within a stromal lamella, and (b) three lamellae with different fibril orientations.

More recent studies found that corneas from other species exhibited significantly different fibrillar collagen structures [59, 60]. The central corneas of horses, cows and marmosets were found to have dominantly superior-inferior (vertical) orientation, while porcine, rabbit and mouse corneas had mainly circumferential orientation. This variation in structure was thought to be due to differences between species in the length, width and insertion position of the rectus muscles, and possibly also due to the existence of the posterior retractor bulbi muscle in all the above mentioned animals but not in humans [61]. However, despite these differences, a common feature that exists in both human and animal corneas is the gradual change in fibril orientation towards a predominantly circumferential organisation close to the limbus.

Other corneal layers include the endothelium; a monolayer of specialised cells that line the posterior surface of the cornea and face the eye's anterior chamber [62]. The endothelium regulates fluid transport across the corneal posterior surface by allowing leakage of solutes and nutrients from the aqueous humor to the cornea and pumping water from the stroma to the aqueous [63]. The small thickness of the endothelium ($\approx$ 2% of corneal thickness), and its cellular structure suggest it has low mechanical stiffness (resistance to deformation) compared to the stroma, although its stiffness has never been quantified [64].

Lining the other surface of the cornea is the epithelium which comprises about 8-10% of corneal thickness [64]. The epithelium functions primarily to block the passage of foreign material into the eye, absorb oxygen and nutrients from the tear film [1] and control the swelling of the stroma [65]. Similar to the endothelium, the epithelium has a cellular structure, which makes its mechanical stiffness low in comparison to that of the stroma.

Between the stroma and the endothelium lies Descemet's membrane, whose physiological function is unknown although it may be to provide mechanical support to the endothelium, or to act as a fluid barrier. Its low mechanical stiffness and small thickness (ranges between 3 and 10 μm or 0.5 to 2% of corneal thickness) suggest it has low contribution to the cornea's overall mechanical behaviour [66, 67].

Finally, Bowman's layer is located between the epithelium and the stroma, composed of strong collagen fibers and has a thickness of 8-12 μm. Similar to Descemet's membrane, the physiological function of Bowman's layer is unknown.

It appears to provide strong attachment of the epithelium to the stroma in human eyes, but the fact that it does not exist in some mammal eyes has not been explained [68]. Earlier research assessed the mechanical behaviour of Bowman's layer and found it to be insignificant [69].

The discussion in this chapter will assess evidence relating the corneal microstructure, and in particular the stromal lamellae organisation, to overall biomechanical behaviour. The implications of this relationship on our understanding of the effects of disease, injury and surgery will also be covered in the conclusion section.

Chapter 3

CORNEAL TOPOGRAPHY

The normal cornea is prolate – steepest at the centre and flattens towards the periphery. A section through the cornea is an ellipse that can be described using Bennett's equation in terms of two parameters; the radius of curvature at corneal apex (apical radius, R_o), and the asphericity or shape factor, p:

$$z = \frac{R_o - \sqrt{R_o^2 - p x^2}}{p} \tag{1}$$

where x and z are coordinates of a general point, a, on the corneal surface, Figure 5. The apical radius, R_o, (which approximately equals the value measured in keratometry) has been measured in a few clinical studies, and examples of the reported average and standard deviation values include 7.87±0.27 and 7.77±0.20 mm [43, 44]. On the other hand, the shape factor, p, provides an indication of how rapidly the surface flattens away from the apex (centre). The average values of p for the human cornea's anterior and posterior surfaces have been found using in-vivo measurements to be 0.82 and 0.62, respectively [43].

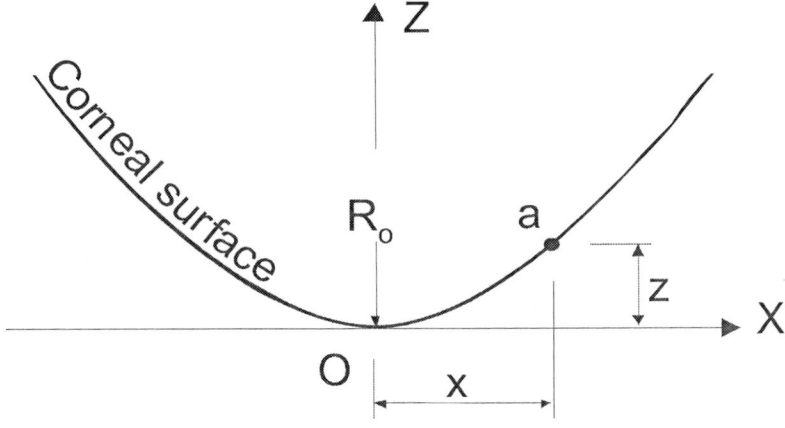

Figure 5. Cross section through a corneal surface with coordinates of a general point a.

CORNEAL THICKNESS DISTRIBUTION

The cornea has a variable thickness that changes from minimum at or close to the centre, to maximum along the limbus, Figure 6. Clinical studies reported different average values, standard deviations and ranges of central corneal thickness, CCT, including 580±54 μm (range = 448-713 μm) [70], 548.2±29.6 (493-617 μm) [71], 535±31 (490-567 μm) [72] and 551.34±40.58 μm for women and 542.34±43.84 μm for men [73].

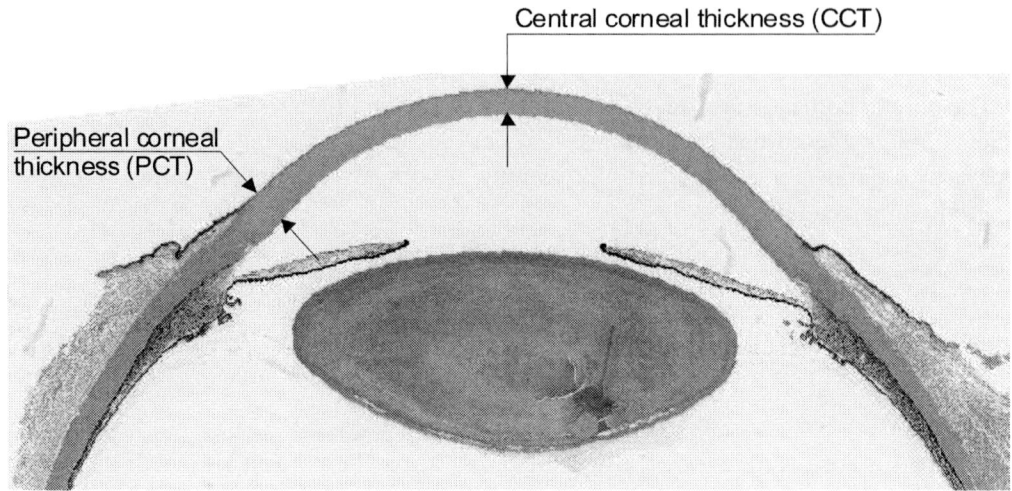

Figure 6. Central and peripheral corneal thickness.

On the other hand, the average values of the peripheral corneal thickness, PCT, reported in the literature varied according to where it had been measured. The values ranged between 600 and 700 μm for measurements taken between 15° to the corneal centre, and at the limbal periphery [74-77]. The relationship between CCT and PCT is yet to be established, although it is known that PCT is usually larger than CCT.

Chapter 4

CORNEAL BIOMECHANICAL BEHAVIOUR

The composite microstructure of the cornea makes its biomechanical behaviour quite complex and poses several difficulties including:

1. With the collagen fibrils being the main load carrying components of the stroma, the behaviour becomes hyperelastic with an initially low stiffness (resistance to deformation) increasing gradually under higher loads.
2. The preferential orientation of stromal lamellae leads to anisotropic behaviour with the mechanical stiffness expected to be highest in the preferred directions of lamellae.
3. The change of lamella orientation from the central area to become circumferential close to the limbus affects the directional anisotropy of the tissue at the transition zone.
4. The cellular composition of the epithelium and endothelium is expected to result in significantly lower stiffness levels compared to the fibrous stroma.
5. The increased corneal thickness away from the centre suggests an associated increase in both flexural and membrane stiffness.
6. Stromal swelling, common during sleeping and following prolonged wear of contact lenses, increases the spacing between collagen fibrils and may affect the biomechanical behaviour of the tissue especially under loads creating flexural actions.
7. The increase in lamella interweaving in anterior stroma, and the reported effect of interweaving in increasing swelling resistance, is expected to lead to variation in biomechanical behaviour between anterior and posterior stroma [78].
8. Diseases such as keratoconus are known to affect the cross linking between collagen fibrils in the stroma and lead to deterioration in corneal stiffness in the affected area. The correlation between the reduced cross-linking in keratoconus and change in mechanical properties requires further study [79].
9. The fast remodelling of the epithelium under external loads, such as those experienced while wearing orthokeratology and other types of rigid contact lenses, makes it difficult to study the cornea's long-term behaviour as in this case epithelial remodelling and corneal creep take place simultaneously.

The discussion below presents the results of recent research studies conducted to characterise corneal tissue and to address some of the challenges described above. In spite of the significant advances that have already been made, current knowledge and understanding of corneal biomechanical behaviour remain imperfect and further progress is needed.

EXPERIMENTAL ASSESSMENT OF CORNEAL BEHAVIOUR

Several experimental studies have been conducted since the 1960s to characterise the biomechanical behaviour of corneal tissue [7, 36, 80-90]. The material constitutive models (stress-strain and Young's modulus-strain relationships) reported in these studies have been compared in Figure 7 and show a wide range of variation [91].

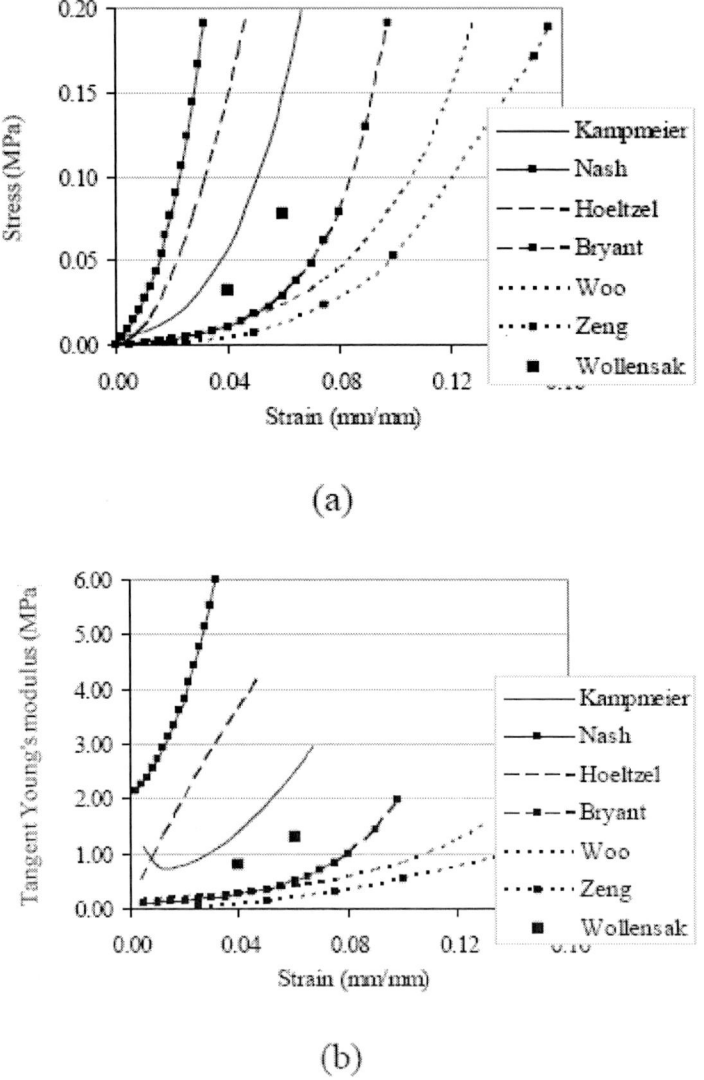

Figure 7. Biomechanical behaviour of corneal tissue as reported in earlier experimental studies, (a) Stress-strain behaviour, (b) Relationships between tangent Young's modulus and strain – Studies reported include Kampmeier [36], Nash [79], Hoeltzel [92], Bryant [7], Woo [87], Zeng [85] and Woolensak [95].

This wide range, which is far beyond what could be acceptable in any predictive analysis of corneal behaviour, is thought to be related to the viscoelasticity and anisotropy of corneal tissue, and the effects of ageing and hydration. Studies adopting different strain rates or employing different testing techniques reported different material properties [79, 85, 88, 92]. Specimens with varying donor ages were sometimes combined in the same test group despite the known age-related changes in the microstructure of corneal tissue (including an increase in stromal fibril diameter [93] and interfibrillar cross-linking [94]) and the associated effect on the tissue's biomechanical behaviour. Corneal specimens with different swelling extents were also used in earlier studies despite the current incomplete understanding of the effect of hydration and swelling on corneal biomechanical behaviour.

For this reason, it has been essential that the characterisation of corneal biomechanical behaviour be based on experimental techniques that offer tight control of strain rate, hydration and temperature, consider the effect of age on tissue stiffness and test the cornea in a state that closely simulates its normal physiologic conditions. Examples of these techniques are described in the next section.

EXPERIMENTAL TECHNIQUES

Two main experimental techniques are currently available to assess corneal behaviour. The first technique, called inflation testing, relies on supporting corneal buttons along the limbus, applying uniform pressure on the cornea's posterior surface, and monitoring the resulting deformation using non-contact methods, Figure 8. The pressure-deformation (p-δ) data obtained experimentally are then converted into stress-strain (σ-ε) behaviour of corneal tissue using shell mathematical analysis through the following steps (see Ref 96 for more details).

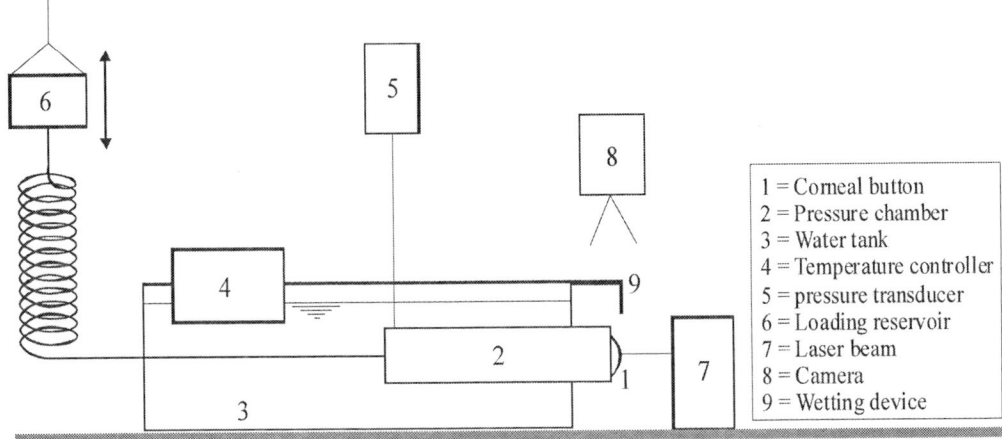

Figure 8. Main components of an inflation test rig.

The secant Young's modulus, E, is first determined for a particular pressure-rise (p-δ) behaviour point as:

$$E = \frac{p\,R_m^2}{2\,\delta\,t}(1-\nu)\left[1 - e^{-\beta\eta}\cos(\beta\eta)\right], \qquad (2)$$

where R_m is the radius of the corneal median surface, t the average thickness assumed = ½ (CCT + PCT), η half the central angle of curvature (see Figure 9), $\eta = \sin^{-1}(R_i/R_{ant})$, R_i radius of the corneo-scleral intersection, R_{ant} radius of the cornea's anterior surface, $\beta = \sqrt{R/t}\cdot\sqrt[4]{3(1-\nu^2)}$, and ν Poisson's ratio, taken as 0.49 assuming corneal tissue behaves as an almost incompressible material [5, 7, 36]. The strain at this behaviour point is defined as:

$$\varepsilon = \frac{p\,R_m}{2\,E\,t}(1-\nu)\left(1 + \nu\,e^{-\beta\eta}\cos\beta\eta\right), \qquad (3)$$

and the stress as:

$$\sigma = \varepsilon\,E. \qquad (4)$$

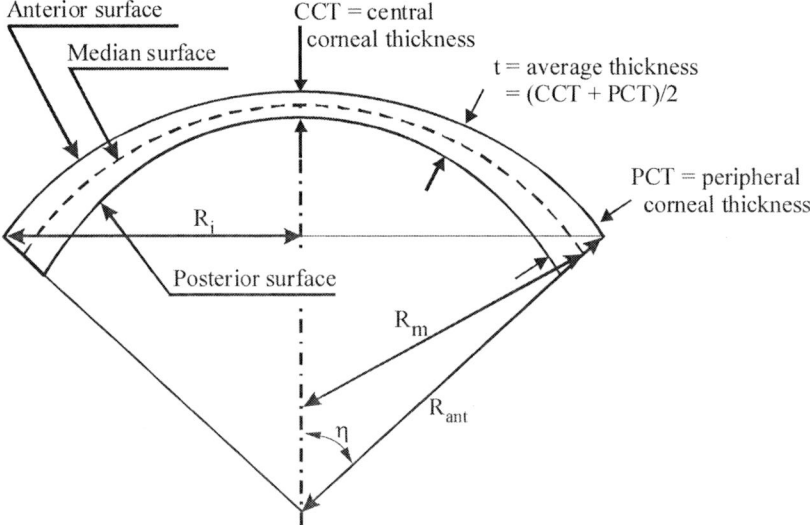

Figure 9. Cross-sectional view of a corneal button showing the main dimensions used in the shell mathematical analysis.

The inflation test maintains the cornea close to its normal working conditions and is therefore considered suitable to determine the global stress-strain behaviour of corneal tissue. The only deviation from normal in vivo conditions is in the development of a pinned edge along the cornea's limbus, which in effect considers the sclera to be stiffer than the cornea to the extent that it can be approximated as a rigid material. This approximation, which is expected to have only a negligible effect on corneal behaviour, has been necessary to make possible the mathematical shell analysis of corneal behaviour, which produced Equations 2-4.

A possible development of inflation testing involves the use of intact eye globes [81], which could be supported along the equator (without preventing radial deformation) and subjected to intraocular pressure loading through a canola inserted into the anterior chamber, Figure 10. However, since no mathematical solution is available to convert the experimentally-obtained pressure-deformation behaviour into stress-strain relationships, the usefulness of this technique is still limited.

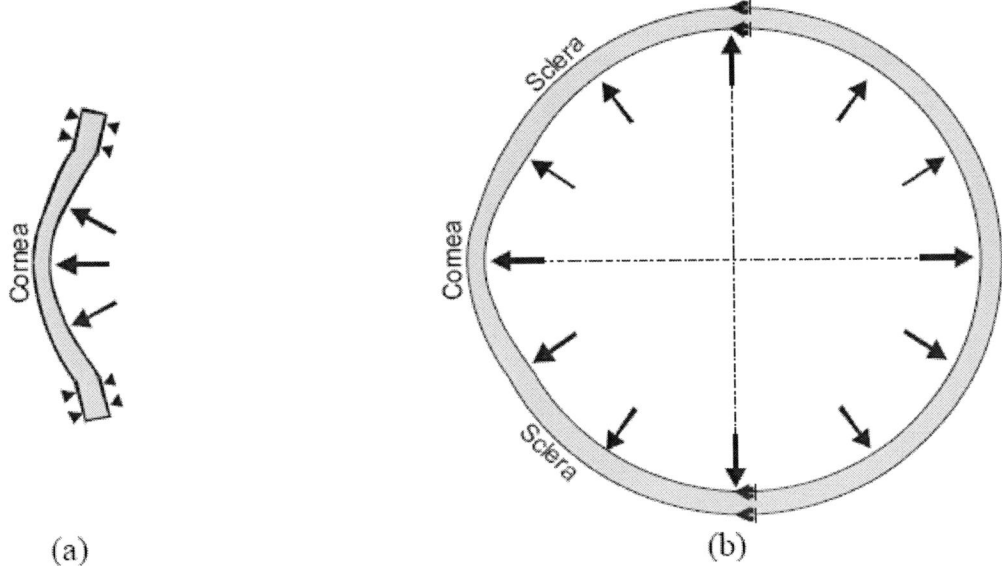

(a) (b)

Figure 10. Loading and boundary conditions applied to (a) corneal button specimens, and (b) whole eye globes subjected to inflation tests.

The second experimental technique is strip extensometry testing, in which a strip of corneal tissue with a constant width is extracted and attached to the grips of a tension machine while monitoring its behaviour. The stresses (σ) and strains (ε) are calculated from the load (T) and elongation (δ) data using the following simple equations:

$$\sigma = T / A \text{ and } \varepsilon = \delta / L, \tag{5}$$

where A and L are the specimen's original cross-sectional area and length, respectively. The simplicity of the technique makes it quite common in ocular biomechanics research but a number of inherent deficiencies, which can reduce its reliability, must be recognised and accounted for. The deficiencies originate mainly from the initially curved form of the specimen (Figure 11), the non-uniform corneal thickness and the termination of fibrils which are not aligned with the specimen orientation.

The straightening of the specimen from its curved form results in initial strains that affect the behaviour under subsequent loading [97], Figure 12. The specimen's curved form also means that its edges are slightly shorter than the longitudinal centre line and this length variation leads in turn to non-uniform distribution of stress under loading. The variable corneal thickness, from a minimum at the centre to a maximum at the ends, further complicates the derivation of the stress-strain relationship from the test load-elongation results. Additionally, the relatively large thickness of the specimen poses another difficulty with the potential of unequal clamping of external and internal corneal layers.

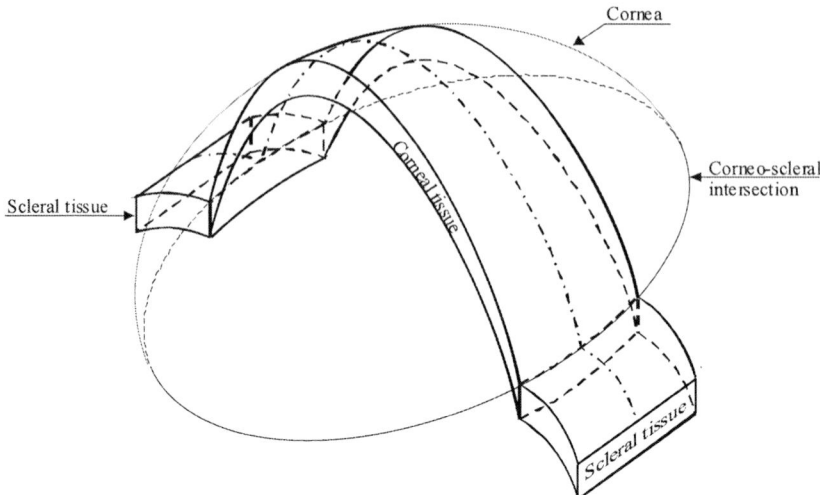

Figure 11. Initial curved form of corneal strip specimens.

Figure 12. Straightening of (a) an initially curved strip specimen produces stresses and strains with the distribution shown in (b).

While these limitations affect the technique's suitability in obtaining the global material properties of the cornea, the technique remains viable for comparative studies such as those on anisotropy (variation in tissue behaviour between different anatomical directions) and viscoelasticity (variation in behaviour with changes in strain rate). In studies on global corneal behaviour, use should be made instead of the above described inflation testing.

However, mathematical techniques have been developed to consider some of the deficiencies of strip testing [98], and these techniques could be useful where strip testing is to be used to produce an approximate estimate of the stress-strain behaviour. The effects of the specimen length variation and the straightening of the initially curved specimen form can be handled first together. Figure 13 shows a 3D view of a strip extracted from a curved cornea (assumed spherical for simplicity) and orthogonal cross-sections showing the variation of specimen length from maximum along the centre line to minimum at the sides.

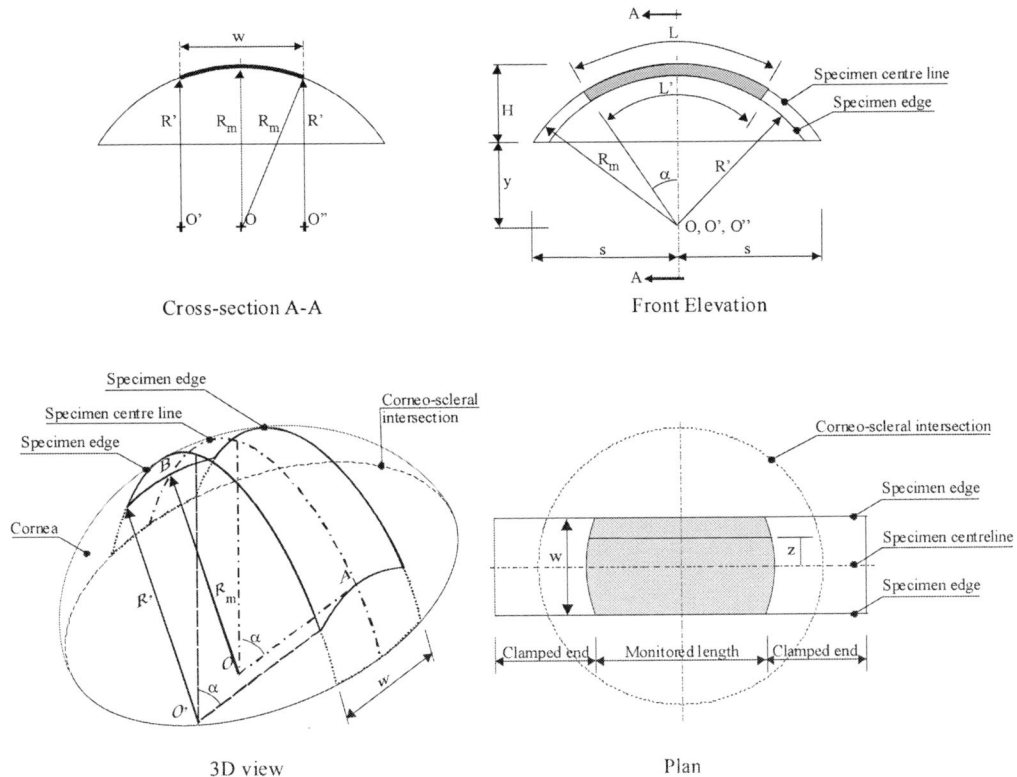

Figure 13. Orthogonal and 3D views of a strip specimen extracted from a corneal button.

Reference is also made to Figure 14 which shows the strain distribution due to (a) the variation in length, (b) initial straightening, and (c) the combined effect. Consideration of these two factors starts with a modified form of the strain, ε (the derivation of Equations 6 to 9 can be found in Ref. 98):

$$\varepsilon = \frac{\delta}{2\alpha \times \sqrt{R_m^2 - z^2}} + \frac{h}{R_m} \qquad (6)$$

where δ is the measured specimen elongation under tension load T, 2α the angle of curvature of the specimen, Figure 13, and R_m the radius of the corneal median surface. Distances h and z are depicted in Figures 13 and 14b, respectively. This form of strain is then used in an equation relating the applied force, T, and the resulting strain, ε:

$$T = \int_{-t/2}^{+t/2} \int_{-w/2}^{+w/2} \left(a\,\varepsilon^3 + b\,\varepsilon^2 + c\,\varepsilon \right) \cdot dz\,dh \qquad (7)$$

where a, b and c are constants to be determined such that the best fit with the experimental data is achieved. With the values of the constants determined, the corresponding stress is then calculated as:

$$\sigma = a\,\varepsilon^3 + b\,\varepsilon^2 + c\,\varepsilon \qquad (8)$$

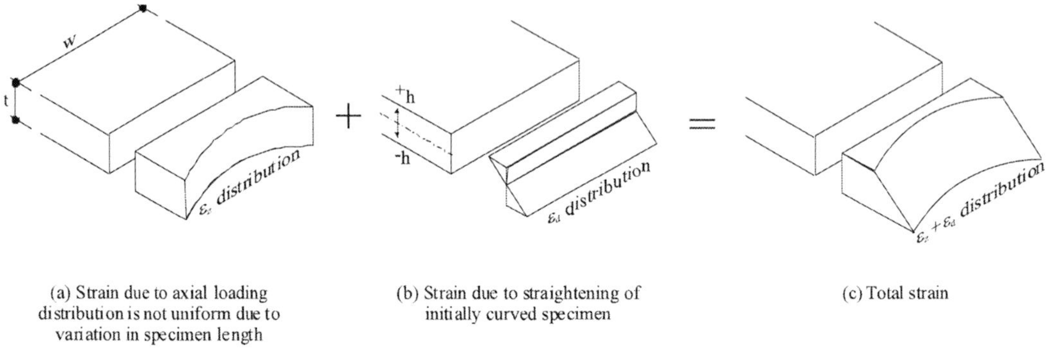

(a) Strain due to axial loading distribution is not uniform due to variation in specimen length

(b) Strain due to straightening of initially curved specimen

(c) Total strain

Figure 14. Strain distribution on the cross-section of a strip specimen.

The effect of thickness variation from minimum at the centre, t_1, to maximum at the clamps, t_2, can be considered next in the stress calculations to develop a modified stress, σ', of the form, Figure 15:

$$\sigma' = \sigma \cdot \frac{t_1}{(t_2 - t_1)} \left[\ln\left(\frac{t_2\,L}{2}\right) - \ln\left(\frac{t_1\,L}{2}\right) \right], \qquad (9)$$

where L is the specimen's initial length, w the specimen width, and σ the stress derived in Equation 8. An assumption of linear variation between t_1 and t_2 is adopted to simplify the development of Equation 9. The second term of the equation:

$$\frac{t_1}{(t_2 - t_1)} \left[\ln\left(\frac{t_2\,L}{2}\right) - \ln\left(\frac{t_1\,L}{2}\right) \right]$$

represents the correction caused by the variable thickness. The second term is below 1.0 and reaches this value only when $t_2 = t_1$.

With these formulations, some of the deficiencies of strip testing could be addressed in the stress and strain calculations.

Other deficiencies such as the termination of some collagen fibrils along the specimen sides and the difficulty in clamping the layered corneal tissue cannot currently be accounted for using similar mathematical techniques.

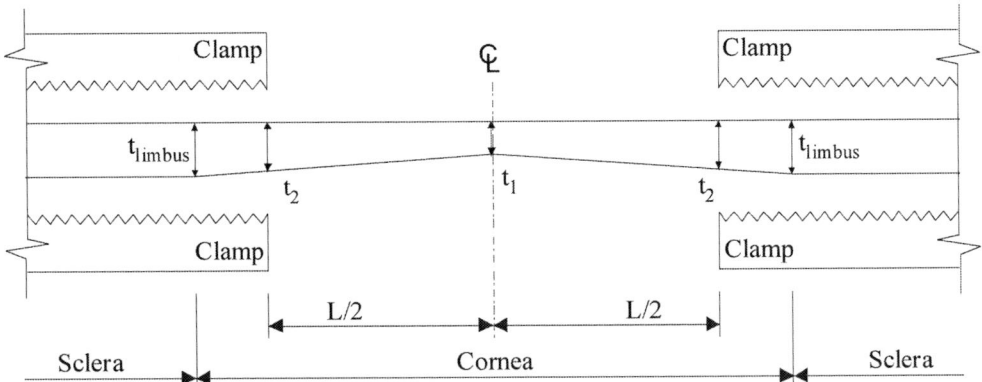

Figure 15. Thickness variation of corneal strip specimens.

INFLATION TEST SETUP

Inflation testing is based on supporting corneal buttons along their limbal tissue, applying uniform posterior pressure that simulates the intraocular pressure and measuring the resulting corneal deformation. The pressure-deformation experimental results are then converted mathematically, using Equations 2-4, to stress-strain tissue behaviour.

An inflation test rig has been developed by the Ocular Biomechanics Group to enable tight control of the conditions thought to have an influence on corneal behaviour including hydration, temperature and load application rate. The rig is totally computer-controlled with all test conditions specified electronically by the user and all behaviour monitoring devices storing their data automatically for later analysis [30].

Fresh and well preserved corneal buttons are mounted onto the custom-built pressure chamber shown in Figure 16 using only mechanical clamps. The pressure chamber is filled either with saline solutions in tests lasting less than 45 minutes, or with the preservation medium Optisol (which can maintain tissue hydration and prevent degradation for up to 14 days post mortem [99-102]) in longer tests. The chamber is connected to a reservoir, which moves vertically up and down using a small motor to control the pressure inside the chamber. The connection between the chamber and the reservoir passes through a large water tank equipped with a temperature controller to set the temperature of the pressure chamber at the required level. The actual pressure in the chamber is measured using an electronic pressure transducer that can be connected to a personal computer. A wetting device is also used to release a slow flow of saline solution on the cornea's anterior surface to prevent its drying.

The motor attached to the reservoir is connected to a variable-voltage power unit that can change the motor speed within a wide range. The ability to change the motor's speed allows the representation of both slow (static) and fast (dynamic) loading conditions to assess the tissue's viscoelasticity. Creep tests can also be conducted by maintaining a constant level of pressure for any reasonable duration of time. Corneal behaviour under loading is monitored using electronic non-contact devices including a laser beam that measures corneal apical deformation with ±1 μm accuracy, and high-resolution digital cameras to record the changes in cornea's profiles caused by pressure changes.

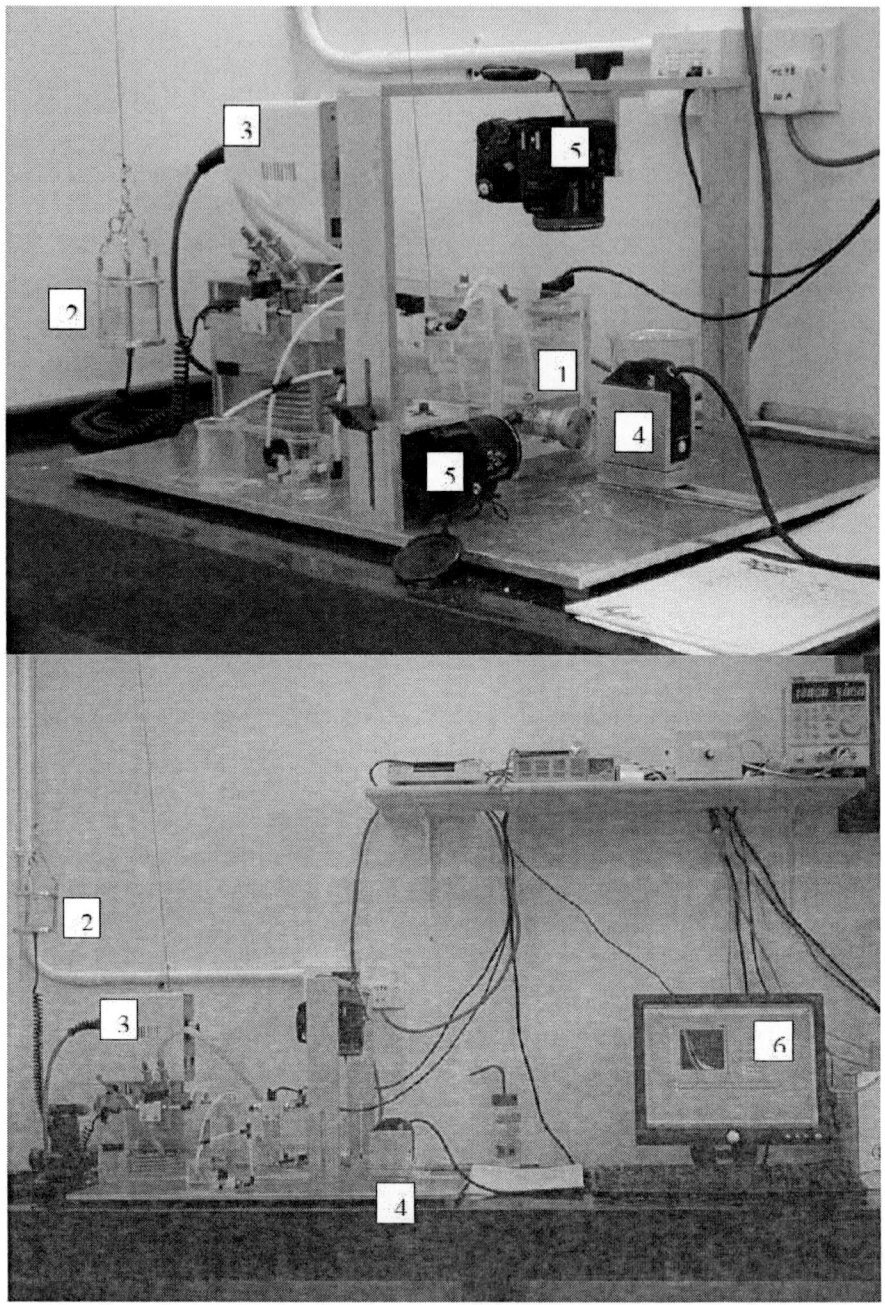

Figure 16. Overall views of inflation test rig showing the main components including (1) the pressure chamber, (2) loading reservoir, (3) temperature controller, (4) laser beam, (5) digital cameras, and (6) control computer.

Controlling the pressure application rate, and the tissue's hydration and temperature, and the use of non-contact monitoring devices have led to experimental results with significantly lower scatter compared to earlier studies which had less control of test conditions. The results obtained using this rig are presented below and used to quantify some of the main biomechanical properties of corneal tissue.

STRIP TEST SETUP

Strip testing is considerably simpler than inflation testing. Tissue strips are extracted from corneal buttons in the desired anatomical direction, connected to mechanical clamps and subjected to uniaxial tension loading in a material testing machine. However, care should be taken to ensure that tissue is fresh and remains well preserved, specimen extraction is in the right location and direction, and connection to the clamps is adequate. Corneal tissue can be preserved in Optisol or Eusol C for up to 14 days post-mortem. Preservation, which also maintains hydration, should continue during the tests as shown in Figure 17. Double blade devices are needed to ensure the uniform width of specimens and the accurate positioning and orientation of the specimen within the corneal button.

Current technology to connect tissue strips to mechanical clamps is inadequate. Too much clamp tightening creates stress concentration in the specimen regions close to the clamps, and can initiate premature failure. On the other hand, insufficient tightening allows slippage and renders the results unreliable.

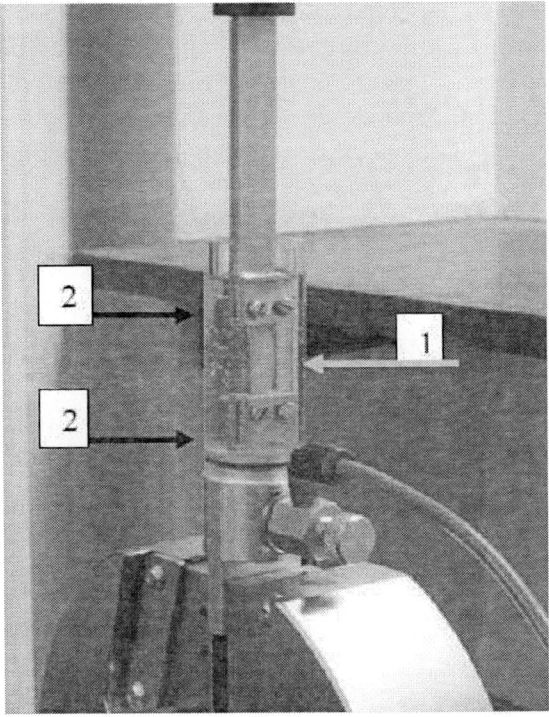

Figure 17. Corneal strip specimen (1) connected to mechanical clamps (2) in a uniaxial tension test.

The fact that the clamps provide non-uniform clamping effect across the thickness of tissue (strong at the outer tissue layers and weaker at mid-thickness) is another cause for concern. However, until a different technique to connect strip specimens to loading machines is developed, mechanical clamps will have to be used with care to minimise the effect of the above described deficiencies.

CORNEAL BEHAVIOUR CHARACTERISTICS

The following sections discuss the corneal behaviour as obtained experimentally using the inflation and strip test rigs described above. In addition to human corneas, the discussion will cover the behaviour of porcine corneas because of their reported biomechanical similarity to, and suitability in some cases as test models for, human corneas. The behaviour characteristics covered include the immediate hyperelasticity, hysteresis and anisotropy of corneal tissue, and the viscoelastic properties represented in the strain-rate effects, and the cornea's creep and stress-relaxation behaviour. Also included are the epithelium's contribution to corneal biomechanics and the effects of the layered stromal construction.

Corneal Hyperelasticity

Under inflation testing, human corneas demonstrate clear hyperelastic (pressure-rise and stress-strain) behaviour with initial low stiffness increasing gradually under higher pressure levels, Figure 18. The end of the initial low stiffness stage coincides with posterior pressure between 15 and 20 mmHg, which is close to the normal intraocular pressure range. The average ratio of final stiffness (observed under IOP higher than 60 mmHg) to initial stiffness (under 5-15 mmHg) has been found close to 11, 9 and 5 for corneas with donors within the age ranges 50-64, 65-79 and 80-95 years, respectively [30].

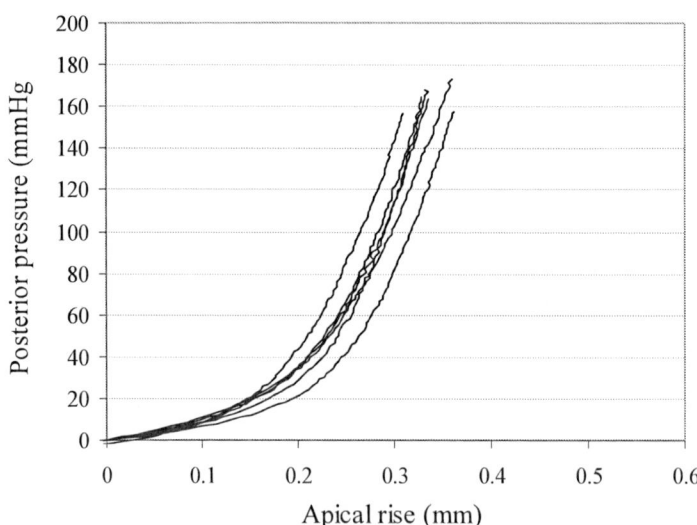

Figure 18. Typical hyperelastic behaviour of human corneas under increasing posterior pressure.

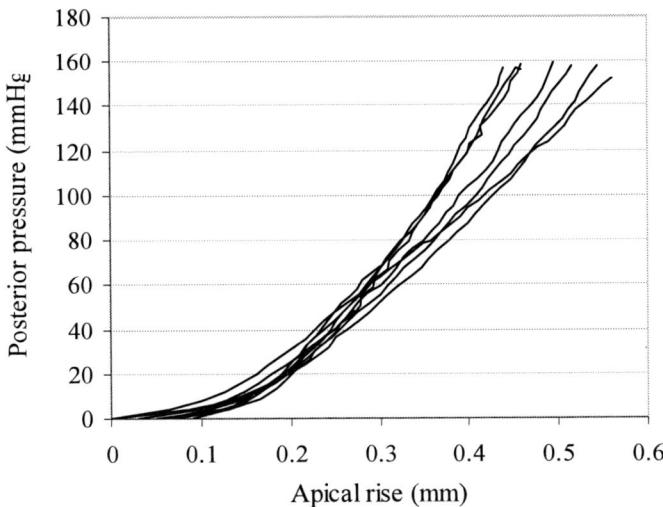

Figure 19. Behaviour of porcine corneas under increasing posterior pressure.

Porcine corneas (aged 4-6 months) exhibit similar hyperelastic behaviour with a number of important differences [35]. The change from the initial low stiffness stage to the final stage with higher stiffness is more sudden and occurs around the lower pressure level of 10 mmHg, Figure 19. The ratio between the final stiffness (under IOP higher than 60 mmHg) and the initial stiffness (under 5-10 mmHg) values is about 5 on average.

Equations with different forms have been tried to fit the stress strain behaviour of human corneas. The best fit was achieved with the exponential equation form $\sigma = A\left(e^{B\varepsilon} - 1\right)$ suggested earlier [7, 87] in which σ and ε are the stress and strain, respectively, and A and B are constants. A clear distinction has been found between the behaviour of specimens within different age groups, and for this reason, three stress-strain equations have been developed of the form:

$$\sigma_{(MPa)} = 0.0070 \left(e^{50\varepsilon} - 1\right) \qquad \text{Age 50-64 years} \qquad (10)$$

$$\sigma_{(MPa)} = 0.0077 \left(e^{59\varepsilon} - 1\right) \qquad \text{Age 65-79 years} \qquad (11)$$

$$\sigma_{(MPa)} = 0.0083 \left(e^{80\varepsilon} - 1\right) \qquad \text{Age 80-95 years} \qquad (12)$$

The ages considered start from 50 years as this was the minimum age of corneal specimens tested. Further work is therefore needed to develop stress-strain relationships for younger human corneas.

The more sudden change in stiffness seen with porcine corneas made it difficult to find an acceptable fit with the above exponential equation form, and an alternative third-order polynomial form was adopted instead:

$$\sigma_{(MPa)} = -350\,\varepsilon^3 + 56\,\varepsilon^2 - 0.21\,\varepsilon \qquad (13)$$

The average porcine behaviour described by this equation is compared in Figure 20 against those of human corneas within the three age ranges considered. Overall, porcine corneas appear to have lower initial and final stiffness values compared with all human corneal groups.

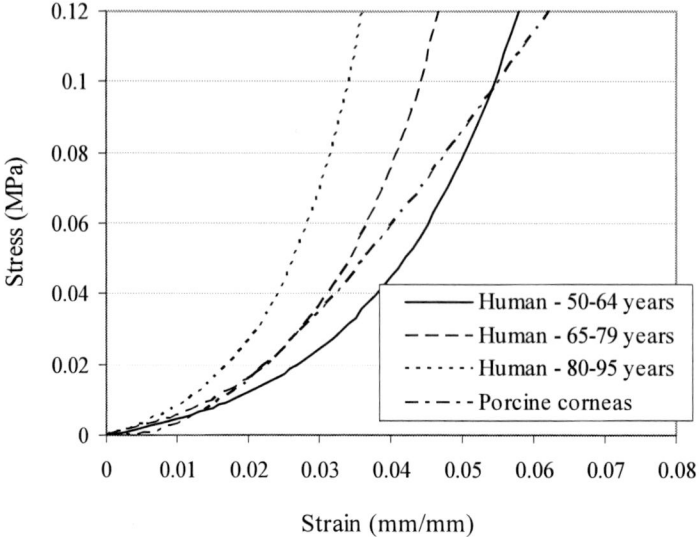

Figure 20. Comparison between the stress-strain behaviour of human and porcine corneas as obtained experimentally using an inflation test rig.

Relationship between Intraocular Pressure and Young's Modulus

The complex hyperelastic form of the cornea's stress-strain behaviour encouraged efforts to find an easier, preferably linear, method to describe the behaviour. One such method is based on presenting the behaviour as relationships between the applied posterior pressure, p, (which represents the intraocular pressure, IOP) and the tangent Young's modulus, E. This method was first attempted by Orssengo and Pye and resulted in the well-known linear relationship [103]:

$$E = 0.0029 \, p \tag{14}$$

Similar relationships have been derived from the experimental pressure-rise results discussed above using a modification of Equation 2 of the form:

$$E_{tangent} = \frac{\Delta p \cdot R^2}{2 \, \Delta r \, t} (1-\nu)\left[1 - e^{-\beta\eta} \cos \beta\eta \right] \tag{15}$$

where Δp and Δr are the increases in pressure and apical corneal rise within a particular pressure increment, and E the instantaneous (tangent) Young's modulus [31].

Corneal Biomechanical Behaviour

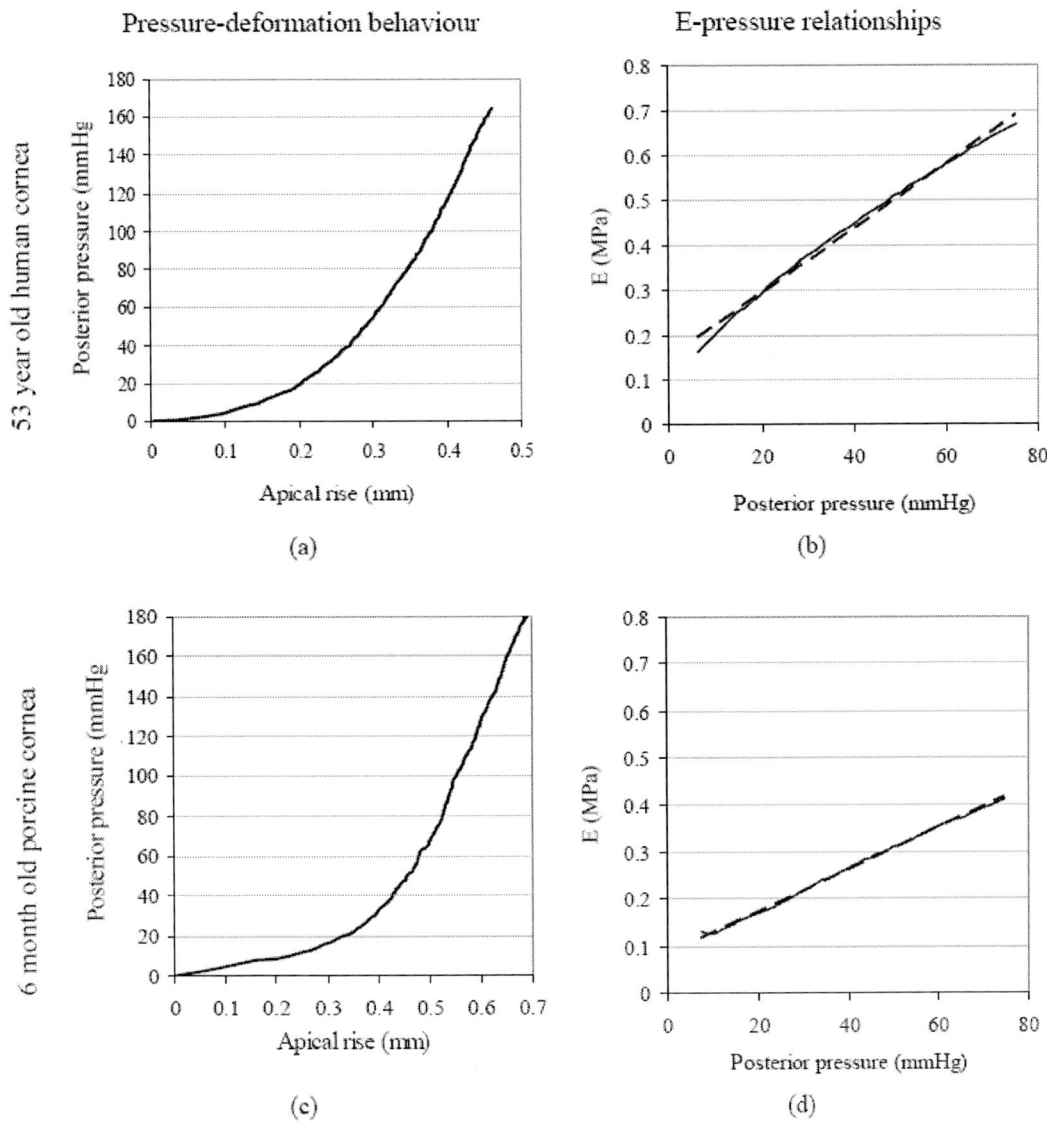

Figure 21. Pressure-deformation and E-p behaviour of example human and porcine corneas, (a,b) 53 year old human cornea, (c,d) porcine cornea.

In spite of the hyperelastic form of both the pressure-rise and stress-strain results, the relationship between Young's modulus, E, and posterior pressure, p, is close to linear. See for example Figure 21 which shows the nonlinear pressure-rise behaviour of a 53 year old human cornea and a porcine cornea, and the corresponding E-p relationships plotted for a pressure range of 0-75 mmHg. In both cases, the E-p relationships are almost linear.

While the E-p results of human specimens are grouped according to the donor age, the porcine results are treated as one group. Within each group, the results are used to find the best fit with an E-p equation of the form: $E = C \cdot p + D$, where C and D are constants. The four equations obtained are:

$E_{(MPa)} = 0.0067\, p_{(mmHg)} + 0.1579$ human corneas, age = 50-64 yrs   (16)

$E_{(MPa)} = 0.0088\, p_{(mmHg)} + 0.1914$ human corneas, age = 65-79 yrs   (17)

$E_{(MPa)} = 0.0107\, p_{(mmHg)} + 0.3527$ human corneas, age = 80-95 yrs   (18)

$E_{(MPa)} = 0.0038\, p_{(mmHg)} + 0.1449$ porcine corneas   (19)

The four equations are compared in Figure 22, which shows porcine corneas with lower E values compared to all human specimen groups. There is also a consistent stiffening effect with age seen in human corneas.

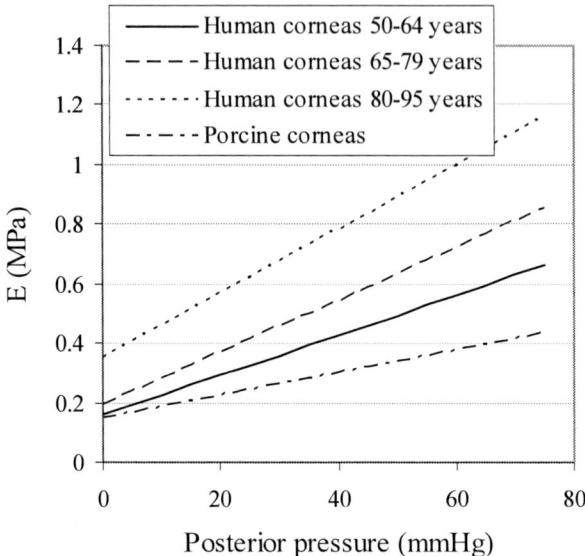

Figure 22. Comparison between E-p relationships for human and porcine corneas.

Corneal Hysteresis

Hysteresis is a viscoelastic property of corneal tissue characterised by the difference in behaviour under loading and unloading conditions, Figure 23. The unloading behaviour has been studied in a similar fashion to the loading behaviour and found to increase in stiffness with age and pressure level. The stress-strain behaviour of the human corneal tissue under unloading conditions can be described using the following equations [34]:

$\sigma_{(MPa)} = 0.00004\left(e^{359\,\varepsilon} - 1\right)$ Age 50-64 years   (20)

$\sigma_{(MPa)} = 0.00005\left(e^{427\,\varepsilon} - 1\right)$ Age 65-79 years   (21)

$\sigma_{(MPa)} = 0.00025\left(e^{510\,\varepsilon} - 1\right)$ Age 80-95 years   (22)

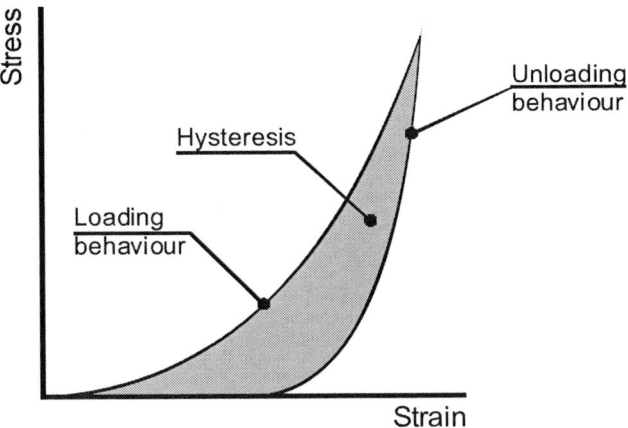

Figure 23. Typical stress-strain behaviour of corneal tissue under both loading and unloading conditions.

An earlier study has considered the unloading behaviour down from different posterior pressure levels, and found the behaviour to be almost independent of the pressure at which unloading starts [34], Figure 24. This particular feature makes the unloading behaviour easy to characterise and implement in the numerical simulation of corneal response.

A direct measure of hysteresis can be obtained experimentally as the difference between the areas under the loading and unloading stress-strain behaviour curves of corneal tissue. Figure 25 depicts experimental results obtained earlier showing a consistent reduction in the magnitude of hysteresis with corneal donor age [34]. The same figure shows a gradual decrease in the areas under both loading and unloading stress-strain curves with age – a finding which can be related to the higher stiffness (and hence lower maximum strain) associated with increased age.

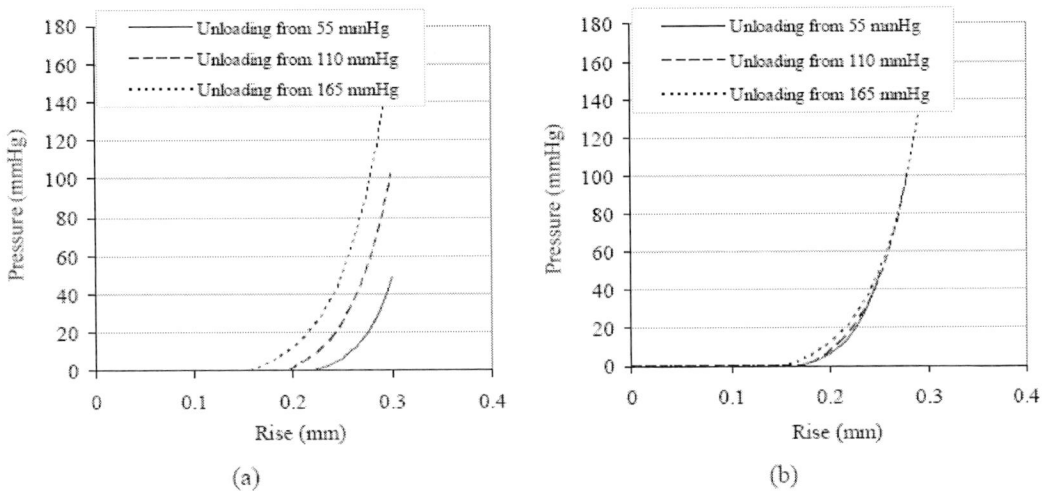

Figure 24. Average unloading behaviour of 80-95 year old corneas, (a) Unloading from three different pressure levels plotted from an arbitrary starting corneal rise of 0.3mm, (b) The three patterns plotted after shifting the behaviour lines with unloading from 55 and 110 mmHg.

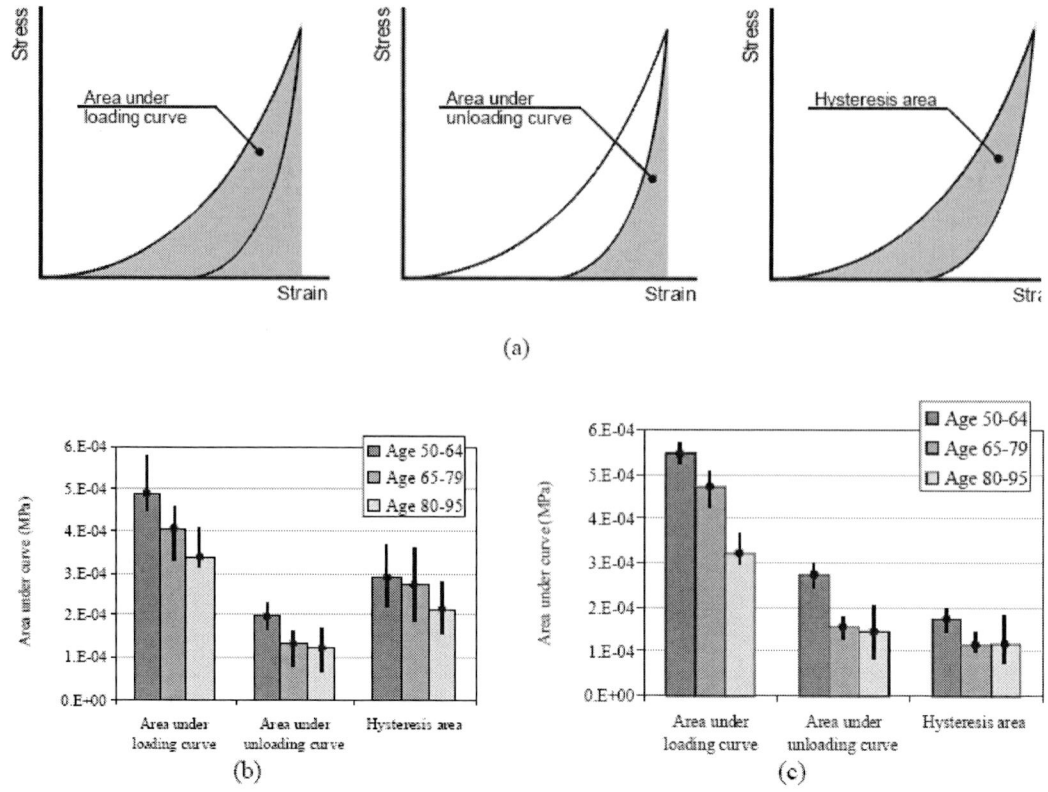

Figure 25. Comparisons between areas under stress-strain loading and unloading curves for corneas with different ages, (a) Sketches showing areas under behaviour curves, (b) Tests under fast pressure application rate of 37.5 mmHg pressure per minute, (c) Tests under slow pressure application rate of 3.75 mmHg pressure per minute – Plots show error bars denoting ranges of variation between specimens within the same group.

Corneal Anisotropy

Strip specimens extracted from human corneas have been tested to determine corneal mechanical anisotropy [28]. The specimens were taken from the vertical, horizontal and diagonal (superotemporal-inferonasal and superonasal-inferotemporal) directions and comparisons of their stress-strain behaviour were held, Figure 26. Vertical specimens were found to be slightly stiffer than horizontal specimens (by 8% on average) and significantly stiffer than diagonal specimens (by 49% on average). This mechanical anisotropy is compatible with the known preferential orientation of collagen fibrils of corneal stroma in the vertical and horizontal directions described in Section 2.

Similar work on porcine corneas found little difference between vertical, horizontal and the two diagonal directions [104]. The differences between the mechanical stiffness of strips taken in the four directions were statistically insignificant and below 3% on average. This apparent mechanical isotropy of behaviour was consistent with the microstructure of porcine corneas in which stromal collagen fibrils were dominantly circumferential, and with no preferential orientation in either the vertical, horizontal or diagonal directions.

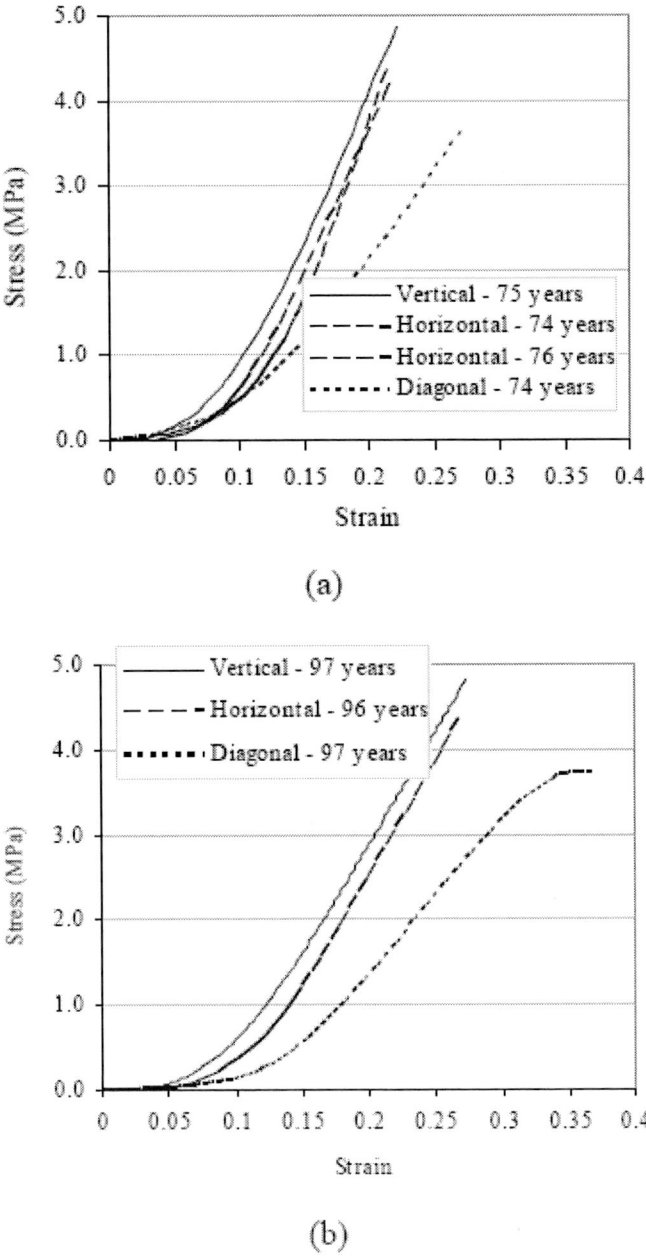

Figure 26. Experimental assessment of the mechanical anisotropy of human donor corneas within different age groups.

Epithelium's Contribution to Corneal Biomechanics

The epithelium is the outermost layer of the cornea, comprising about 10% of its thickness. It functions primarily to block the passage of foreign material into the eye, absorb oxygen and nutrients from the tear film [1] and control the swelling of the stroma [65]. Little work has been done to quantify the epithelium's contribution to corneal biomechanics, although researchers have tended to discard the epithelium's contribution as insignificant [7, 92, 105].

Recent research used experimental testing of human corneas before and after removing the epithelium and found that, on average, the epithelium was approximately 10% the stiffness of stromal lamellae of the same thickness [38].

In order to compare the two sets of pressure-rise relationships (with and without the epithelium), the two lines with the largest and smallest rises were identified within each set and plotted together to mark the boundaries of behaviour seen with and without the epithelium.

These lines were compared in Figure 27a, which shows little difference between the two groups, although specimens with intact epithelium appeared to be slightly stiffer (i.e. experiencing less deformation for the same pressure) than the other group. More precise comparisons were obtained by considering the average stiffness values at specific pressure levels between 10 and 100 mmHg. In this case, stiffness was defined as the posterior pressure in mmHg required to produce an increase in corneal central rise of 1 mm. The stiffness comparisons shown in Figure 27b demonstrate consistently higher average stiffness values associated with the specimens with intact epithelium.

However, the ratios between corresponding values remained within 1-3% in all cases, and these differences were not statistically significant (P>0.05 from t-tests) due to the considerable overlap of the test pressure-rise results. With intact corneas exhibiting higher stiffness by 1-3% than corneas without the epithelium, and with the epithelium being about 10% the corneal thickness, the stiffness of the epithelium is estimated between 9 and 27% that of the stroma.

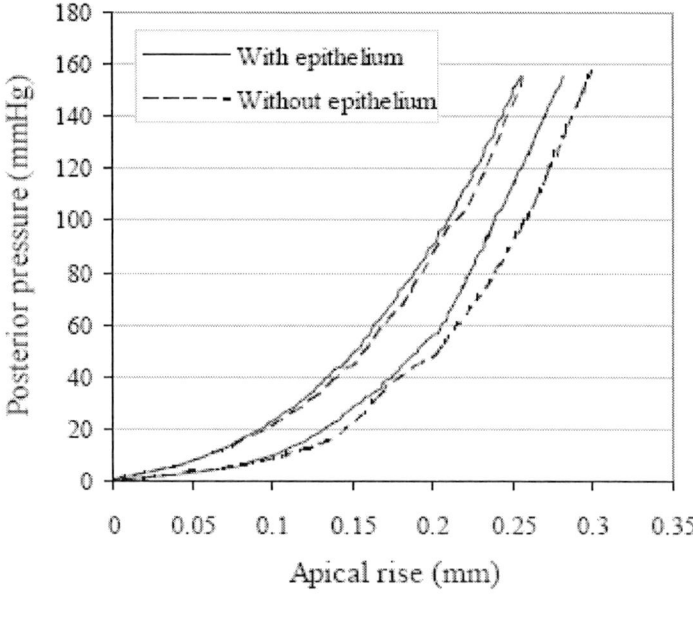

(a)

Figure 27. (Continues)

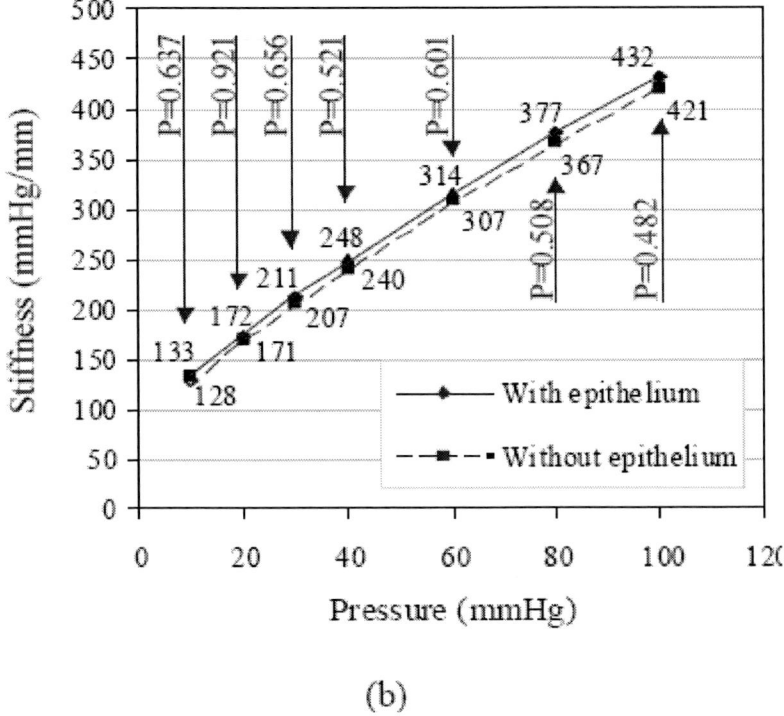

(b)

Figure 27. Comparison of pressure-rise behaviour between corneas with and without the epithelium – Results of statistical analysis of differences between the two specimen groups are included on figure in the form of t-test P values, (a) Range of pressure-rise results for all specimens tested, (b) Average values of stiffness calculated at specific levels of posterior pressure.

Stromal Inter-Lamellar Adhesion

The corneal main layer, the stroma, is analogous to a stack of 300 sheets of copy paper that have some ability to slide relative to each other if subjected to shear [9]. Lack of knowledge of the shear behaviour affecting the relative movement of stromal sheets (or lamellae) led to the exclusion of this feature from numerical simulations [106, 107]. Earlier simulations concentrated on global behaviour, in which it was possible to assume homogenous biomechanical properties (with no consideration of inter-lamellar shear) derived from tests involving intact corneas. These modelling approaches would not be suitable for the simulation of keratoconus for instance, with its abnormal inter-lamellar cross-links [108, 109], and refractive surgery procedures involving the severing of anterior lamellae.

A number of earlier studies determined through experimental means the adhesive tensile strength of the connection between stromal lamellae [40-42]. Strip specimens extracted from human stromal tissue were split at their ends to break the connection between two stromal lamellae, Figure 28. The ends were then clamped and pulled apart, effectively subjecting the interface between stromal lamellae to direct tension. This test, conducted on 68 human specimens resulted in inter-lamellar tensile strength values between 0.069 and 0.077 MPa.

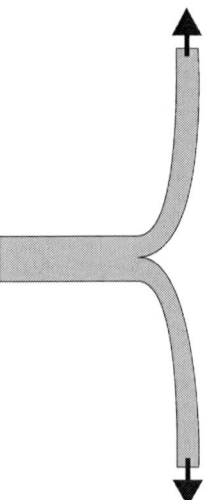

Figure 28. Testing of corneal tissue to determine the adhesive tensile strength of the connection between stromal lamellae.

Further tests subjecting square specimens of stromal tissue to direct shear quantified the shear behaviour of stromal tissue [39]. The behaviour was found to be nonlinear and had the following exponential form:

$$\tau = 0.009 \left(e^{1.74 \gamma} - 1 \right) \qquad (23)$$

where τ and γ were the shear stress and shear strain, respectively. This result allowed an indirect assessment of the assumption of stromal isotropy. If isotropy is maintained, the behaviour patterns under shear and surface tension would have been interrelated. This possibility is checked in a simple test, in which an approximate value of shear modulus, G, of 0.03 MPa was obtained from the shear stress-strain Equation 23 (0.02 MPa shear stress divided by 0.65 strain). With an assumed value of Poisson's ratio, ν, of 0.49 (based on considering corneal tissue an incompressible material [7]), the material's Young's modulus, E, would be estimated as E = G [2 (1 + ν)] = 0.09 MPa. This value is considerably lower than those reported in earlier experimental studies [30], indicating that the assumption of stromal isotropy was invalid.

Corneal Long-Term Behaviour

Behaviour under long-term loading can be determined using two main tests, namely; creep and stress-relaxation. In creep tests, specimens are kept for a period of time under a constant level of load or stress while monitoring the gradual increase in deformation. On the other hand, stress-relaxation tests involve subjecting specimens to a certain load, then maintaining the resulting deformation for a period of time while monitoring the gradual decline in stresses, Figure 29.

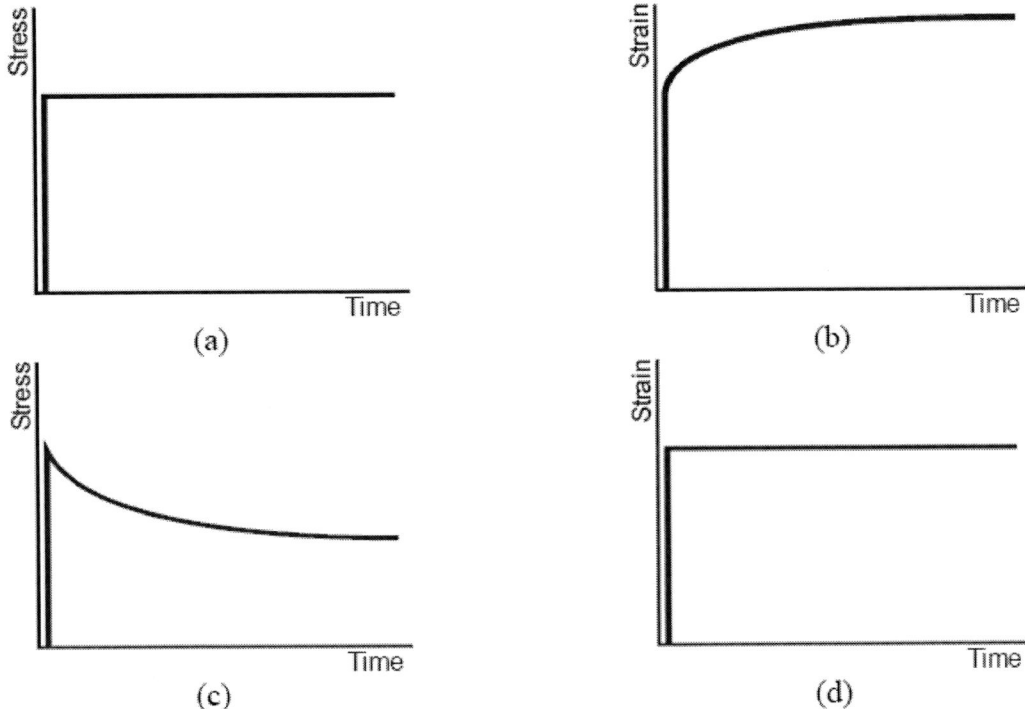

Figure 29. Overview of the stress and strain progression in (a,b) creep and (c,d) stress-relaxation tests.

The creep behaviour of human corneas attached to the inflation test rig and subjected to a constant posterior pressure was found to differ with age as shown in Figure 30 and to follow Equations 24-26 [35]:

$$\varepsilon_{creep} = 16 \times 10^{-6} \left(e^{2.28 \, [\log t - 1]} - 1 \right) \quad \text{age} = 50\text{-}64 \text{ yrs} \quad (24)$$

$$\varepsilon_{creep} = 16 \times 10^{-6} \left(e^{2.23 \, [\log t - 1]} - 1 \right) \quad \text{age} = 65\text{-}79 \text{ yrs} \quad (25)$$

$$\varepsilon_{creep} = 18 \times 10^{-6} \left(e^{2.11 \, [\log t - 1]} - 1 \right) \quad \text{age} = 80\text{-}95 \text{ yrs} \quad (26)$$

where t is the creep time in seconds. A similar equation was developed for porcine corneas of the form:

$$\varepsilon_{creep} = 5 \times 10^{-6} \left(e^{2.02 \, [\log t - 1]} - 1 \right) \quad \text{porcine corneas} \quad (27)$$

Figure 30 demonstrates a gradual decrease in the average human corneal creep with age, which is more evident between the ages 50-64 and 65-79 years. However, there is a notable overlap of results as demonstrated by the error bars in the figure. A statistical analysis of the results using t-test shows that the association between increased age and reduced creep strain is insignificant.

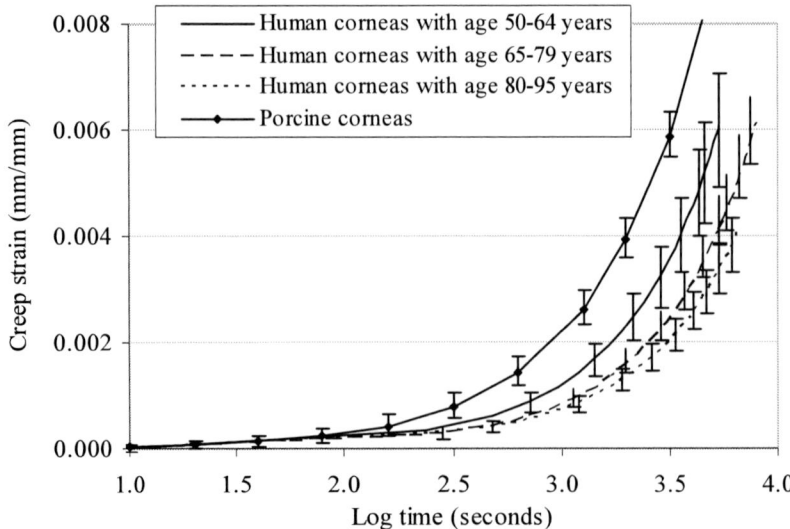

Figure 30. Comparison between average creep behaviour of porcine corneas and human corneas with different ages.

In the mean time, porcine corneas undergo significantly more creep strain compared to all human corneas. The ratio of average creep strain experienced by human and porcine corneas at 300 and 1000 seconds is 49% and 42%, respectively. Figure 30 also shows that the overlap between porcine and human creep-time results is limited to the first stages of the test, up to about 100 seconds, beyond which there is clear separation. A statistical analysis of the two sets of data further returned significant association between creep results and specimen type with P < 0.001.

Stress-relaxation tests were also conducted, in which corneal strips were held with the length reached under a uniaxial load of 4N, then 8N and 12N. The sustained load, and hence the stress, were observed to reduce gradually with time. The stress reduction, or relaxation, was highest at the start of the test and reduced gradually with time. The stress-time results obtained in an earlier study were as shown in Figures 31a,c for human and porcine corneas, respectively [35]. The results show a consistent trend in which porcine corneas were observed to lose more of their initial stresses than human corneas. This trend was evident under all three values of the initial load.

Dividing the stresses in each test by the initial stress value under either 4N, 8N or 12N, and obtaining the average behaviour pattern in each case resulted in the plots in Figures 31b,d. Displaying the results in this form reveals two trends including an evidence that porcine corneas relax quicker than human corneas. After 400, 800 and 1200 seconds, porcine corneas respectively lose 49%, 55% and 59% of their initial stresses on average. The corresponding average stress reductions in human corneas are 28%, 30% and 32%. The differences in relaxation rates between human and porcine corneas are statistically significant (P<0.001). Further, the stress-relaxation behaviour appears to be largely independent of the initial applied stress. The error bars displayed in Figures 31b,d show significant overlap between the results obtained under different initial applied loads.

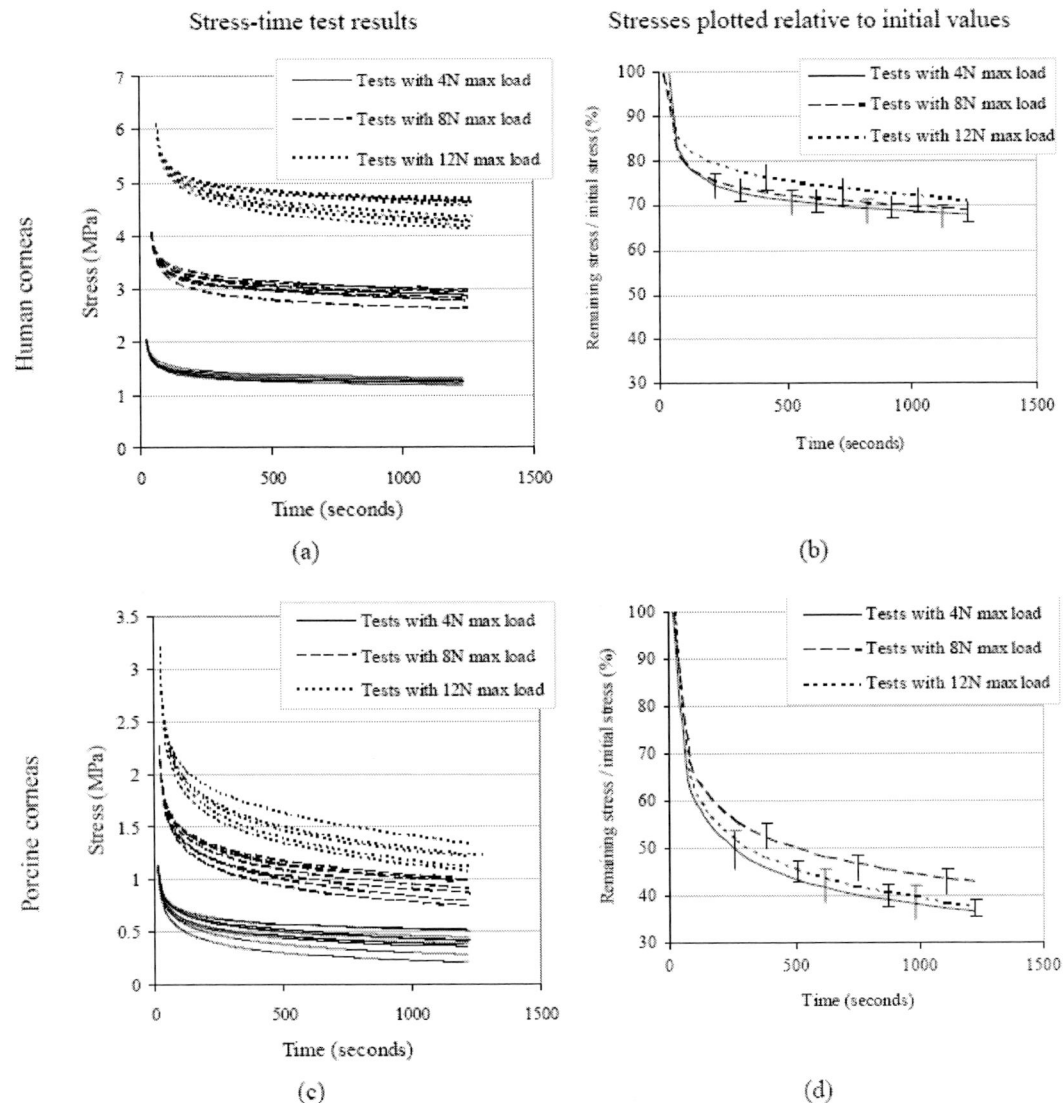

Figure 31. Stress relaxation behaviour of human and porcine corneas, (a,b) Human corneal behaviour, (c,d) Porcine corneal behaviour, (a,c) Stress-time experimental results, (b,d) Results plotted in relation to initial stress values.

The average trend lines are also close and do not show any consistent trends related to the value of initial load. This characteristic simplifies the development of stress-relaxation behaviour patterns for use in numerical simulation applications.

Corneal Dependence on Strain-Rate

As a viscoelastic material, the behaviour of corneal tissue is time-dependent and is expected to vary with the rate at which the material is strained or loaded. Experimental tests involving vertical strips of human and porcine corneal tissue have been conducted under uniaxial tension and with varying strain rates. With specimen length of 12 mm, width of 4 mm and elongation rates between 0.1 and 25 mm/minute, the strain rate varied between 0.008 and 2.1 per minute. As shown in Figure 32, the stiffening experienced with increasing the strain rate was significant at the slow end of the rate range and became progressively less notable as the strain rate increased. Under slow loading, increasing the strain rate from 0.008 to 0.08 per minute caused an increase in material stiffness of 47% and 9% on average in human and porcine corneas, respectively. In contrast, the corresponding average increases in stiffness that corresponded to increasing the strain rate from 0.83 to 2.1 were 1.5% and 1.2%. The consistent difference in the response of human and porcine corneal tissue to strain rate changes remains unexplained.

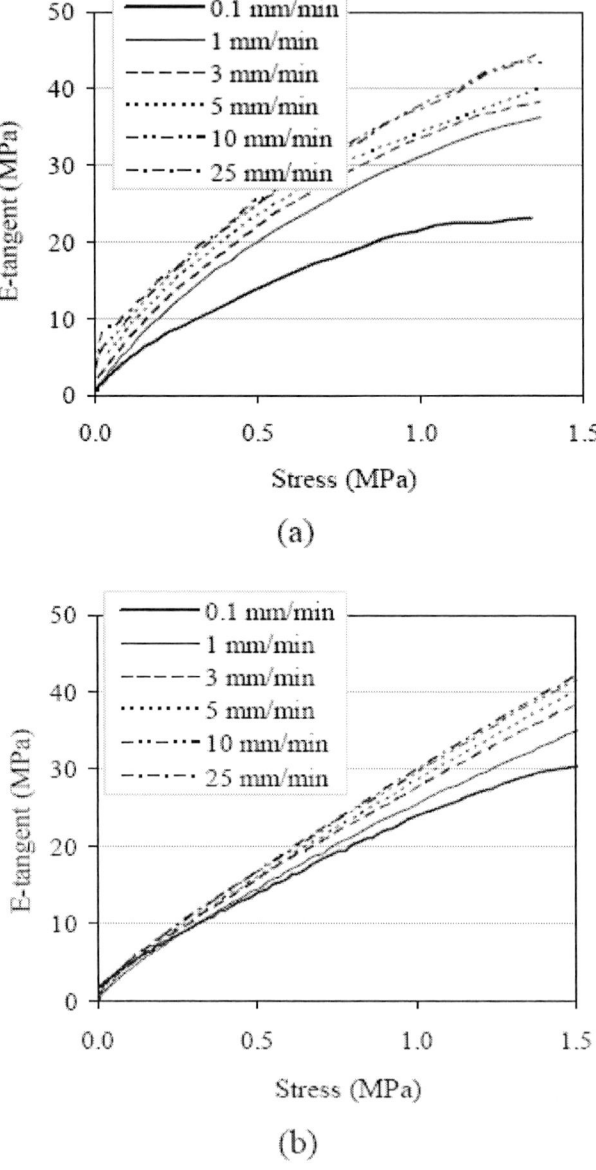

Figure 32. Effect of strain rate on the behaviour of (a) a human, and (b) a porcine corneal strip specimen tested under uniaxial tension.

Concluding Remarks on Corneal Material Behaviour

Characterisation of the biomechanical properties of corneal tissue is key to developing proper understanding of overall ocular behaviour, and response to disease, injury and surgery. It is also essential for the construction of accurate predictive numerical simulations of ocular behaviour, which have several clinical applications including improving the accuracy of tonometry, the planning of refractive surgery procedures, the design of rigid contact lenses, the treatment of myopia and keratoconus, and guiding surgical interventions following eye injuries.

The experimental studies reported above led to fast progress in our understanding of the human cornea's biomechanical behaviour, but several gaps still exist in our knowledge. Overall hyperelastic and hysteretic behaviour, and how it changes with age, has been characterised, but as the corneas tested were aged 50 years or over, similar studies are needed to consider younger corneas. The epithelium's contribution to corneal biomechanics has been quantified and further work is needed to assess the endothelium's and Bowman layer's contributions, and to distinguish between the anterior and posterior parts of the stroma (which differ in lamella interweaving). The correlation between corneal microstructure and biomechanical behaviour has been established in normal healthy eyes, and similar correlations are needed in swollen eyes (e.g. due to diurnal effects) and in eyes with keratoconus. Further work is also needed to build similar knowledge in the biomechanical behaviour and microstructure of the sclera, without which it becomes difficult to understand overall ocular behaviour or to construct predictive numerical models of whole eye globes.

It is also important to have the ability to measure corneal and scleral biomechanical properties in vivo. This is necessary if we are to build accurate and reliable patient-specific ocular numerical simulations, especially with the notable variations in biomechanical behaviour observed experimentally in same-age donor corneas. Recently, Reichert produced the Ocular Response Analyzer (ORA), a tonometer that produces a hysteresis parameter and claims that this parameter provides a measure of corneal biomechanics [110, 111]. However, the true biomechanical origin of the hysteresis measurement, and which biomechanical property it represents, is not yet fully understood.

Overall, and in spite of the current gaps in our knowledge, it remains true that the significant progress made to date in corneal material characterisation, topography mapping and microstructure detailing is offering an unprecedented opportunity to construct accurate predictive numerical simulations of corneal response to disease, injury and surgery. This subject will be discussed in some detail in the remainder of this chapter.

Chapter 5

NUMERICAL MODELLING OF OCULAR BIOMECHANICS

In general, a numerical model is a computer-based model that describes the physical behaviour of a structure and is able to predict its response to external loading. The structure is divided into a large number of simple parts called elements, and the behaviour of each element is described in a set of equilibrium equations normally presented in a matrix form. The matrices are then assembled together to form overall equilibrium equations for the structure. The mathematical assembly of individual matrices is done in the same way as the elements can physically be put together to form the actual structure. Solving the overall equilibrium equations under any set of given loads then provides a prediction of the structure's response to the loads including its deformation and stress distribution [112]. This modelling and solution process is commonly described as finite element analysis.

An example corneal model is shown in Figure 33. The model has four element layers including anterior and posterior layers representing the epithelium and endothelium, respectively, and two internal layers representing the stroma. Each layer is divided into 2904 solid prismatic elements, each with 5 faces and 6 corner points, called nodes. The elements are arranged in 6 radial segments and 22 circumferential rings in such a way that ensures all element internal angles remain between 20° and 70°. The model has a thickness that varies from 540 μm at the centre to 690 μm along the edge, central anterior radius of 7.8 mm, anterior shape factor, p = 0.82, and corneo-scleral diameter of 11.5 mm.

For finite element (FE) models to be accurate and predictive of real behaviour, the models need to closely represent the structure's geometry, topography, boundary conditions, loading regimes and material properties. In FE ocular models, these conditions include the non-uniform corneal thickness, the elliptical topography, the connection to the sclera, the uniform effect caused by the IOP, and the hyperelastic, hysteretic, anisotropic, viscoelastic and layered behaviour of corneal tissue. However, since the introduction of these properties can make the construction of FE models quite complex and the analysis costly and time consuming, it becomes important to distinguish between the properties with significant effect on model predictions and those that complicate the analysis without important improvements in accuracy.

It is well known that all FE models involve simplifications in topography, material properties, boundary conditions and/or loads [113-115].

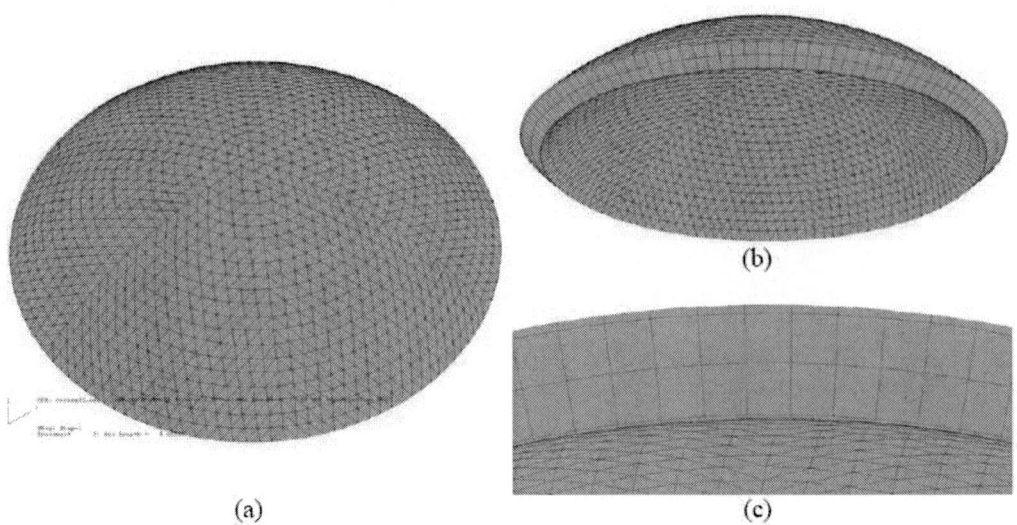

(a)   (b)   (c)

Figure 33. Views of a corneal numerical model with 4 layers, each with 2,904 elements arranged in 6 segments and 22 element rings, (a) Top (anterior) surface of model showing model segments and element rings, (b) View of bottom (posterior) surface and thickness of model, (c) Cross-sectional view showing the 4 layers of the model representing the epithelium, stroma and endothelium.

For instance, the elliptical corneal topography is sometimes approximated as a spherical surface, or the hyperelastic material behaviour idealised using a linear-elastic model. Some simplifications are essential as they reduce the complexity of the model, lower its development and execution costs and make it easier to modify and adapt. However, the more simplifications adopted in FE model construction, the farther away from reality the model becomes, eventually affecting its reliability. It is therefore important in constructing FE models that the user adopts simplifications where they are likely to have a negligible effect and avoid them when they start to erode the accuracy of predictions. This exercise requires experience and validation against known experimental behaviour.

The following discussion aims to assess the effect of simplifications in FE model construction with particular attention to simplifications in corneal topography, material behaviour and boundary conditions.

The discussion refers to numerical results for a corneal model under two simple load cases; a uniform posterior pressure (simulating the intraocular pressure, IOP) and a concentric anterior point load, and the simplifications in a particular parameter are assessed by comparing their effect on the simulation results. This exercise is followed by an overall discussion on how to improve the accuracy of ocular numerical models and to allow them to have a more useful role in clinical practice.

The two loading cases considered have been chosen for their simplicity and because they represent idealised situations where the cornea is subjected to either uniform membrane stresses under posterior pressure, or localised bending stresses under point loads. They have been preferred to subjecting the cornea to more realistic loadings such as tonometric pressure or external impacts since these loads create combinations of membrane and bending stresses, which would make the discussion of results less straightforward. The analysis used the model described above and shown in Figure 33, with some modifications added in the study of each simplification as detailed below.

For simplicity, the following discussion focuses on the deformation recorded at the centre point of the anterior surface, described as the corneal central, or apical, rise. Stiffness as used in the discussion of results means corneal resistance to deformation, and in particular central corneal rise.

SIMULATION OF CORNEAL NON-UNIFORM THICKNESS

The effect of corneal thickness variation is studied by considering two models; one with variable thickness; from 540 μm at the centre to 690 μm along the edge, and one with constant (average) thickness of 615 μm, Figure 34a. The results shown in Figures 34b, c demonstrate a notable effect of thickness variation in models under point loads. The stiffness increases by 10% on average when adopting a constant thickness of 615 μm, rather than variable thickness. The effect under uniform posterior pressure is considerably lower, and limited to 1.6%.

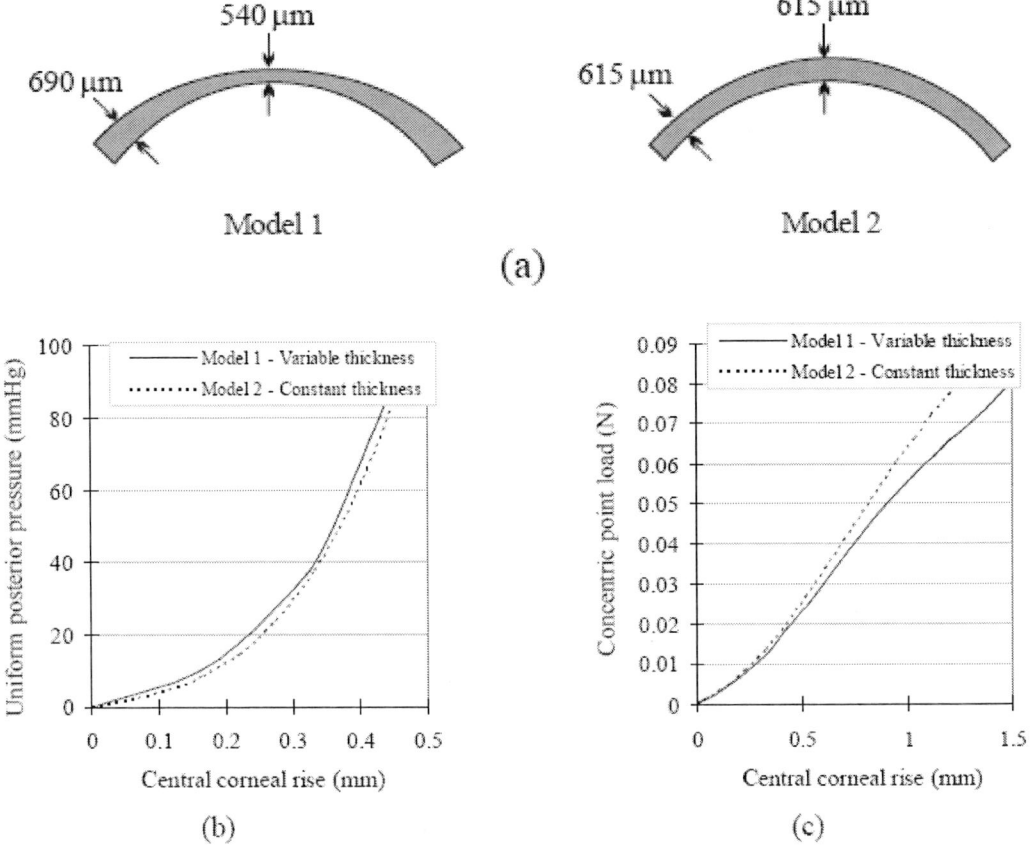

Figure 34. Behaviour predictions obtained using two models (a); one with variable corneal thickness and one with constant thickness, under (b) a uniform posterior pressure, and (c) a concentric point load.

The structural effects of the two loading conditions differ considerably. In the first loading condition, point loads create highly localised bending stresses around the load application point, and this concentration of stress makes the behaviour dependent on the thickness under the load, and almost independent of the thickness variation elsewhere. On the other hand, the behaviour under the tension stresses caused by posterior pressure depends on the thickness in all parts of the corneal surface, and in this case, replacing the real-life variable thickness with an average constant thickness is not expected to lead to notable change in overall corneal behaviour.

SIMULATION OF CORNEAL TOPOGRAPHY

The cornea has a prolate elliptical profile giving it maximum curvature (or smallest radius) at the centre and flatter surface towards the limbus. The effect of elliptical profile is studied under both uniform posterior pressure and anterior point load, and the results are depicted in Figure 35. Models with shape factors, p, between 0.6 and 1.2 are considered to cover the average reported prolate value, 0.8, the spherical case, 1.0, and an oblate case, 1.2 [43-45]. The results show a small, but notable, effect on models subjected to point loads (max effect = 6.3%) and a more evident effect, exceeding 28%, under posterior pressure.

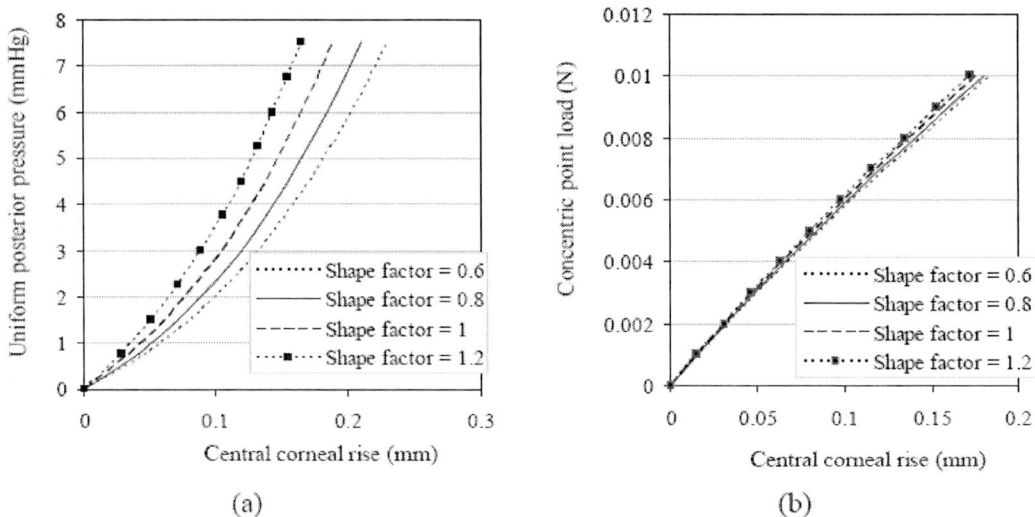

Figure 35. Effect of corneal shape factor, p, on predicted behaviour under (a) a uniform posterior pressure, and (b) a concentric point load.

The highly localised stresses created by point loads are mainly dependent on the thickness under the load, and corneal topography away from the load application point has little effect. On the other hand, the stresses developed in the cornea due to posterior pressure depend on the topography. When the corneal median surface matches the catenary topography, the cornea experiences only tension membrane stresses (i.e. no bending stresses) and hence undergoes small deformations. However, as the cornea deviates from the catenary topography, bending stresses with increasing values arise leading to increased deformation and reduced stiffness.

SIMULATION OF CORNEAL BOUNDARY CONDITIONS

The connection to the sclera presents a problem in corneal simulation as the sclera is not stiff enough to prevent all deformation along corneal edge and is not flexible enough to allow free edge movement.

This makes the use of pinned supports along corneal edge unrepresentative of real conditions. One possible technique to avoid this problem is to use whole ocular models (involving at least both the cornea and the sclera), but while this solution presents a close representation of the cornea's natural conditions, it enlarges the model significantly and makes it expensive to build and analyse.

Another solution is to provide the cornea with titled supports to consider the expected sclera deformation, Figure 36a.

In this case, the slope of the supports is controlled such that the deformations of the cornea-only model and the whole-globe model almost match. The results of this exercise are demonstrated in Figure 36.

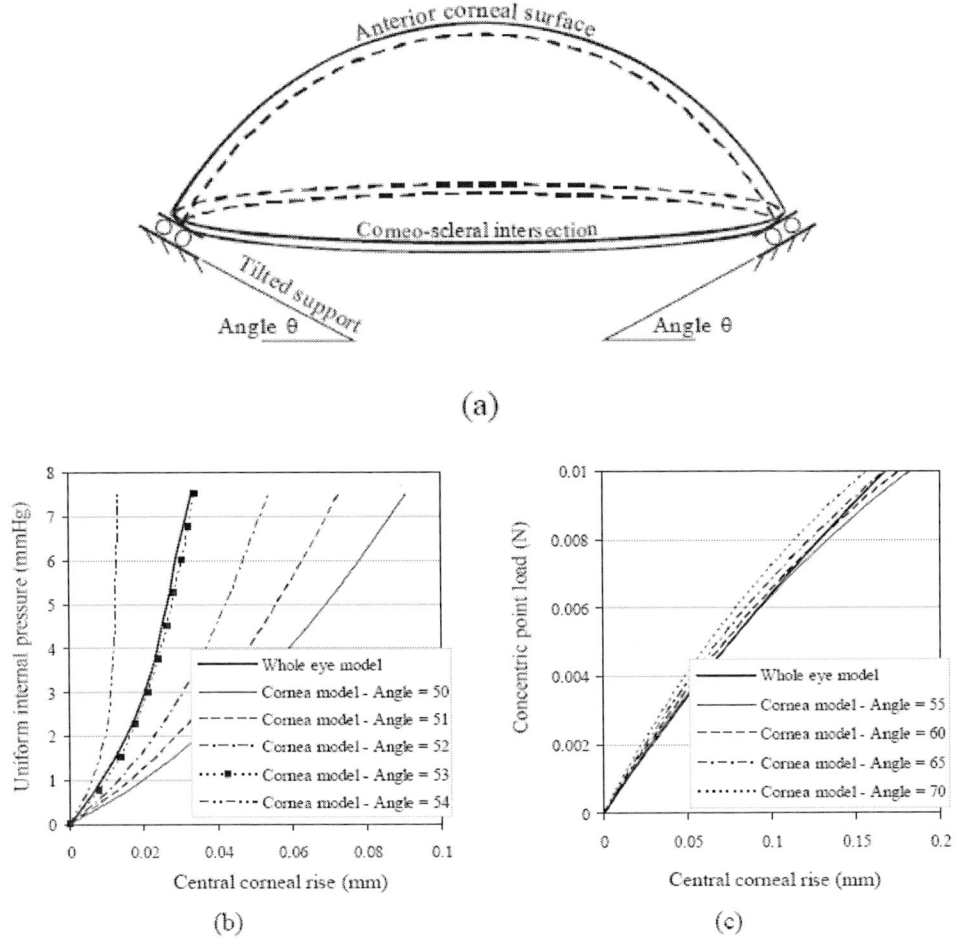

Figure 36. Behaviour of (a) cornea-only models with titled supports under (b) a uniform posterior pressure, and (c) a concentric point load.

The results show first that the support tilts suitable for simulations under posterior pressure and point load were 53° and 60°, respectively. Second, as the behaviour under point loads is dependent mainly on the thickness under the loads and largely insensitive to model changes elsewhere, it is reasonable to expect the model's low sensitivity to variations in support tilt in this case. The results also show the high sensitivity of analysis under posterior pressure to the support tilt. This finding should be considered carefully as it can make cornea-only models in this case less attractive, and much less reliable, than whole eye models.

SIMULATION OF CORNEAL HYPERELASTIC BEHAVIOUR

Corneal tissue is known to possess hyperelastic behaviour with initial low stiffness, gradually increasing under higher stress levels. Considering this behaviour pattern is a challenge since the engineering materials commonly considered in commercial FE packages have the opposite behaviour, i.e. initial high stiffness, decreasing under higher stresses. For this reason, several earlier modelling studies simplified the corneal behaviour into a linear-elastic pattern, and this was justified by the assumption that common loads on the cornea create low stresses. In Figure 37, this assumption is tested under both posterior pressure and a central point load, and in both cases, two models were considered with hyperelastic and linear-elastic material behaviour patterns, respectively. The hyperelastic model considered the following stress-strain relationship developed earlier for corneas aged 50-65 years [30]:

$$\sigma_{(MPa)} = 0.0070 \left(e^{50\varepsilon} - 1\right), \tag{28}$$

while the linear-elastic model had a constant Young's modulus, E, equal to the E of the hyperelastic model under low strain, $\varepsilon = 0.01$ (E = 0.56 MPa). The results show that the linear elastic model provided reasonable results under low load levels up to about 0.002 MPa (15 mmHg) or 0.01 N, beyond which the deviation between the two models became unacceptably high.

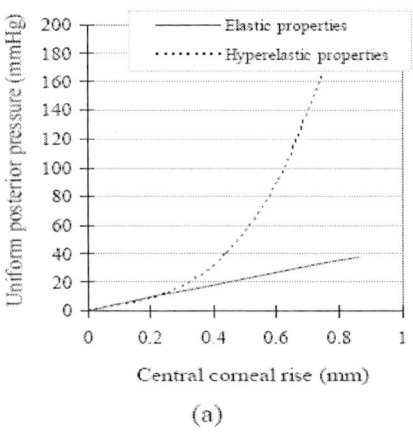
(a)

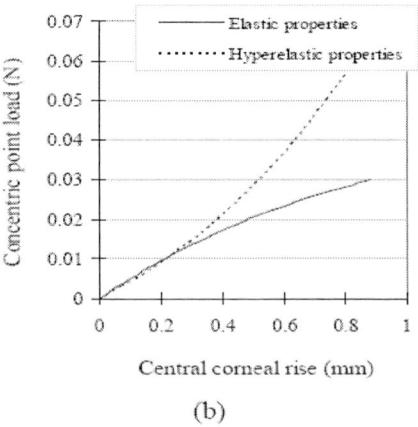
(b)

Figure 37. Effect of adopting a linear-elastic material behaviour pattern on the model's response to (a) a uniform posterior pressure, and (b) a concentric point load.

SIMULATION OF CORNEAL LAYERED CONSTRUCTION

The cornea's three main layers, the epithelium, stroma and endothelium, have different material properties. The stroma, with its fibrous structure, has the highest stiffness, while the cellular epithelium and endothelium are much more flexible. An earlier study estimated the epithelial stiffness at 10% that of the stroma [38]. No study has been carried out on endothelial biomechanics, but its cellular structure could justify assuming it had similar properties to the epithelium. The effect of considering the distinctive material properties of each layer on corneal behaviour was assessed and the results are shown in Figure 38.

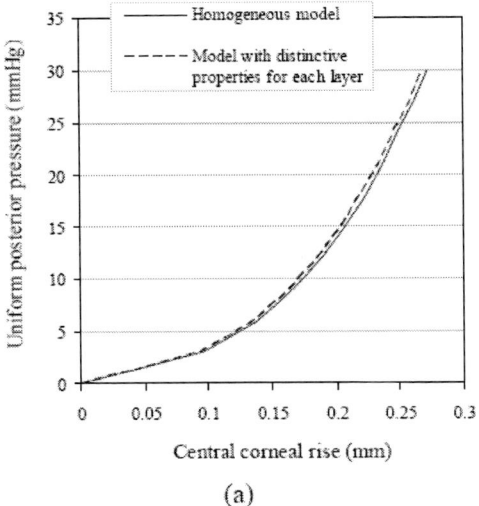
(a)

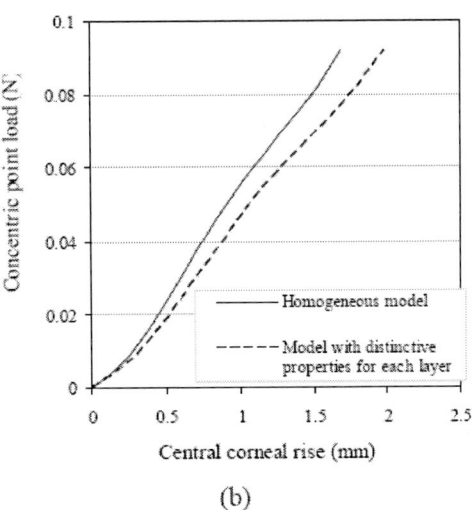
(b)

Figure 38. Effect of considering the distinctive material properties of the main corneal layers on predicting the behaviour under (a) a uniform posterior pressure, and (b) a concentric point load.

The homogenous model had a uniform stiffness equivalent to that of the heterogeneous model under loads creating membrane tension. This condition meant that:

$$(E \cdot t)_{homogenous} = (E \cdot t)_{epithelium} + (E \cdot t)_{stroma} + (E \cdot t)_{endothelium} \qquad (29)$$

In this relationship, corneal thickness, t, was split into 8%, 90% and 2% parts for the three layers, respectively [64]. $E_{epithelium}$ and $E_{endothelium}$ were also both taken as 10% of E_{stroma}. The results show almost negligible effect of varying the material properties on the behaviour under posterior pressure, as would be expected. However, the effect was clear under a concentric point load with an average reduction in stiffness of 15%.

Since posterior pressure creates mainly membrane tension stresses, these stresses will be shared by the cornea's three layers according to their respective membrane stiffness values (E.t). However, as the overall membrane stiffness of the cornea does not change by varying the distribution of stiffness between layers (according to Equation 29), the overall behaviour under posterior pressure is not expected to change.

On the other hand, the bending stresses caused by point loads depend on the cornea's bending stiffness calculated as the sum of $E_i \cdot I_i$ for all corneal layers, where I_i (the second moment of area of layer i) is $t_i^3 / 12 + t_i \times z_i^2$, and z_i the distance between the mid-thickness of layer i and the cornea mid-thickness. Therefore, placing the layers with high Young's modulus, E, nearer the cornea mid-thickness (hence lower z_i), as in the second model, reduces the overall bending stiffness and leads to increased deformation.

SIMULATION OF STROMAL INTER-LAMELLAR ADHESION

The cornea has weak shear stiffness at the surfaces separating stromal lamellae [9]. As was discussed above, the stromal inter-lamellar adhesion was assessed experimentally, and the shear stress-strain, τ–γ, relationship $\tau\,(\text{MPa}) = 0.009\left(e^{1.74\gamma} - 1\right)$ was derived [39]. The effect of stromal adhesion on corneal behaviour was considered by reducing the stromal inter-layer contact in the numerical simulations from full to the level observed experimentally. The results depicted in Figure 39 show a consistent trend in which corneal stiffness degraded with reducing stromal inter-lamellar interaction. On average, the stiffness reduced by 6% and 56% for the cases under posterior pressure and point load, respectively.

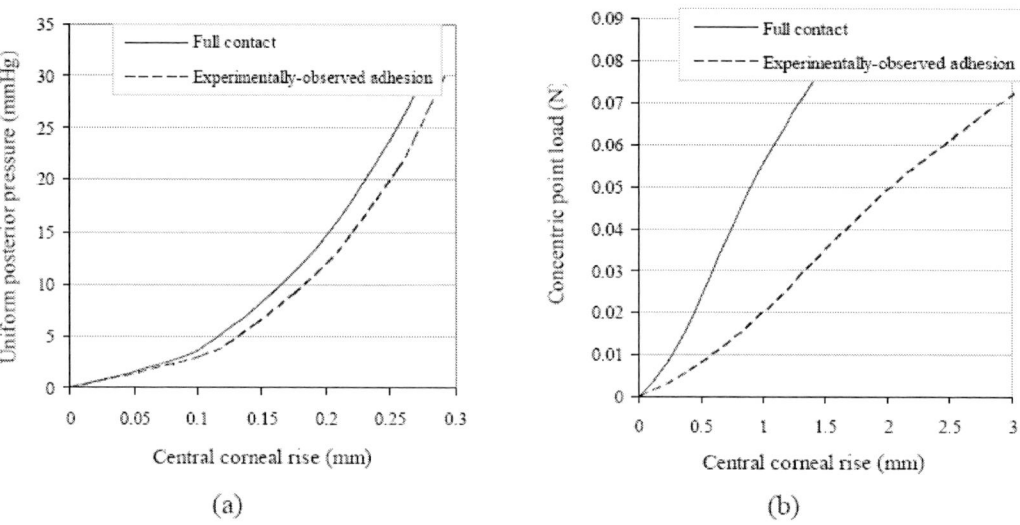

Figure 39. Effect of stromal inter-laminar adhesion on corneal behaviour under (a) a uniform posterior pressure, and (b) a concentric point load.

Since posterior pressure is transferred through membrane tension, which depends on the thickness and stiffness of individual layers, and not on inter-layer contact, reducing the level of contact does not have a considerable effect on overall behaviour under pressure [116]. On the other hand, reducing inter-layer contact reduces the second moment of area (as the stiffness term $t_i \times z_i^2$ is to be multiplied by a reduction factor which depends on the level of interaction) and therefore affects the resistance to deformation under bending actions such as those created by point loads.

SIMULATION OF CORNEAL ANISOTROPY

The human cornea has a preferential orientation of stromal collagen fibrils in the vertical (superior-inferior, SI) and horizontal (temporal-nasal, TN) directions, leading to anisotropy in biomechanical behaviour [24, 26]. The degree by which the SI and TN directions exhibit higher stiffness than diagonal directions has been determined earlier [28] and discussed above. Implementing anisotropy in corneal simulations was studied by considering two multi-layer numerical models; one with anisotropic and another with isotropic behaviour, respectively. The models in this study had 8 element layers representing the stroma, instead of the two layers used in other parts of the study.

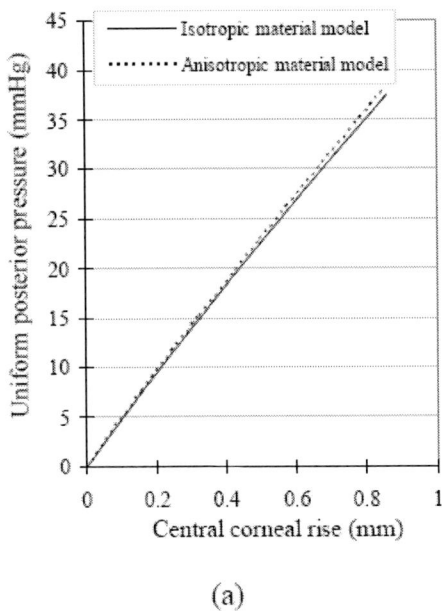 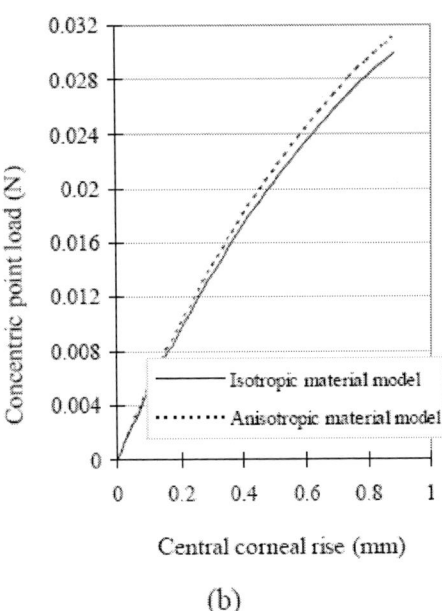

(a) (b)

Figure 40. Effect of introducing material anisotropy on predicted behaviour under (a) a uniform posterior pressure, and (b) a concentric point load.

In the first model, odd-numbered stroma layers were assumed to have fibrils in only the vertical direction while even-numbered layers had fibrils in the horizontal direction. The layers were given a high stiffness in the fibril direction and 10% this value in the perpendicular direction.

In the second isotropic model, the stiffness was the same in all directions and taken as the average of the stiffness values in the two directions of the first model. The models were used to predict the behaviour under a posterior pressure and a concentric point load and the results showed little effect, below 2% under posterior pressure, and 4% under point load, caused by introducing anisotropy, see Figure 40.

SIMULATION OF CORNEAL HYSTERESIS

Considering corneal hysteresis in the material definition in FE models is perhaps the most difficult and challenging feature in model construction. This feature requires that the corneal material is defined with both loading and unloading behaviour patterns, and that the simulation can select the pattern that fits the stress history of each element. Following each change in loading, the simulation calculates the stress level within each element and determines whether the element is to undergo further loading or unloading, and based on this decision, the element is allowed to follow the more suitable behaviour pattern from the stress level determined earlier, Figure 41. This process is difficult to program and makes the analysis process quite complex.

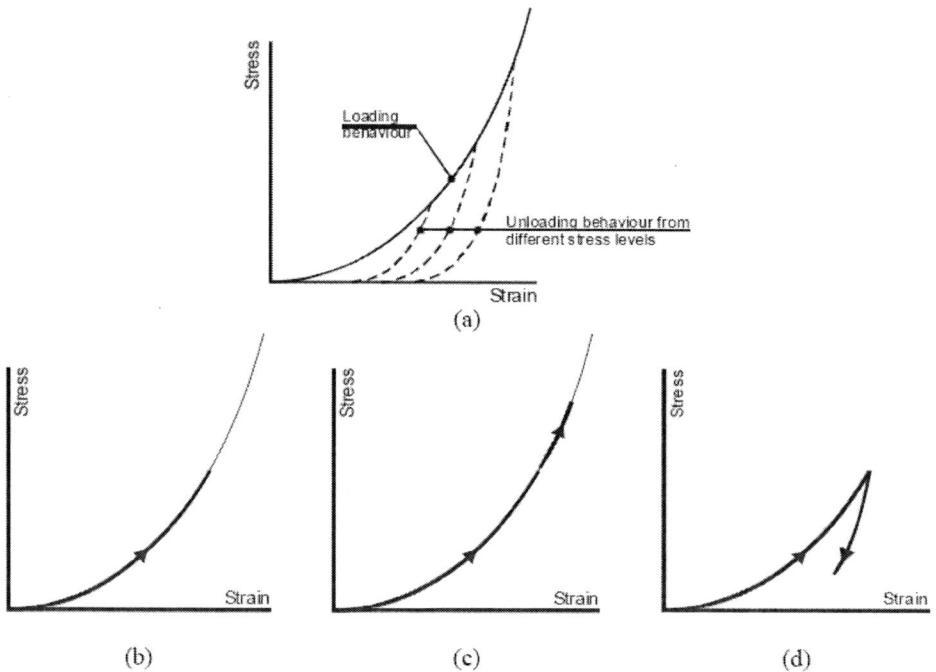

Figure 41. (a) Behaviour of corneal tissue under both loading and unloading conditions, and the stress development in a particular element during an FE analysis including (b) the stress history, and the behaviour adopted if the element is to undergo either (c) further loading or (d) unloading.

Nevertheless, an assessment is needed of whether hysteresis should be incorporated in material definition in the first place. The first trials under monotonically increasing posterior pressure or point loads showed no effect of considering hysteresis. Since all elements underwent further loading in all analysis steps, consideration of the unloading behaviour was not required.

However, further tests showed different outcomes. First, a simple corneal model was subjected to a posterior pressure that increased gradually to 30 mmHg then decreased down to 0. Ignoring hysteresis resulted in pressure-rise predictions that were the same under loading and unloading, Figure 42a. However, when hysteresis was considered, the unloading behaviour differed from the loading behaviour as would be expected in experimental tests, Figure 42b.

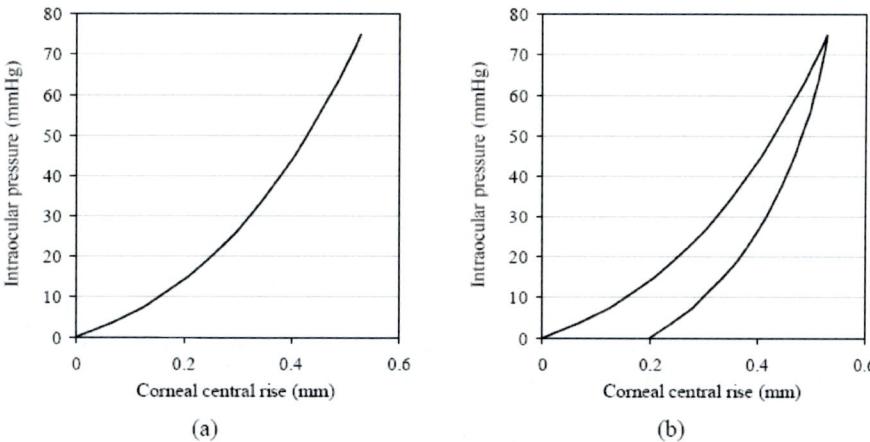

Figure 42. Model predictions of the behaviour of a cornea subjected to a cycle of pressure increase and decrease while (a) ignoring and (b) considering the unloading behaviour of corneal tissue.

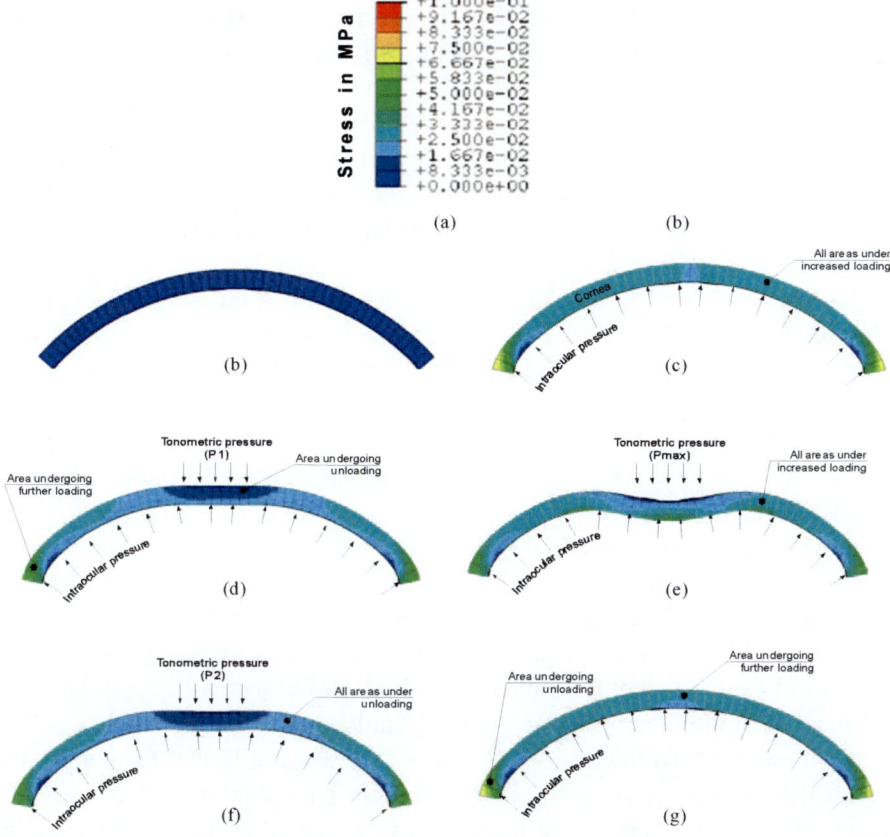

Figure 43. Stress distribution diagrams for a cornea with CCT = 540 μm, PCT = 690 μm, R = 7.8 mm, p = 0.82, age = 57 years and IOP = 15 mmHg, during the progress of the ORA procedure showing areas of loading and unloading, (a) Key showing stress values in MPa, (b) Cross-sectional view at start, (c) Stresses after applying the intraocular pressure on the model's posterior side, (d) Stresses at first applanation, (e) Stresses under Pmax, beyond which the air pressure starts to decrease, (f) Stresses at second applanation, (g) Stresses after complete removal of air pressure.

Another test involved the simulation of the Ocular Response Analyzer (ORA) – a tonometric technique that applies an air pulse on the cornea in increasing then decreasing intensity. Figures 43b-g show the stress distributions on a corneal cross-section at different stages of loading; first by intraocular pressure and then by external air pulse. The figures illustrate the unloading experienced in the middle region of the model during the second stage (force-out) of the ORA procedure. It is therefore hypothesised that this region should follow the material unloading behaviour in order to trace corneal response during the full ORA procedure.

In order to test this hypothesis, the ORA procedure was simulated using two models that considered and ignored corneal hysteresis, respectively. The models were subjected first to an intraocular pressure of 15 mmHg. The ORA air pressure was then gradually applied until applanation was detected at P1 (the first applanation pressure) of 21.3 mmHg. The pressure was further increased to a peak value, Pmax, of 53 mmHg according to the relation: Pmax = 1.1713 P1 + 28.106 [117]. After reaching Pmax, the pressure was decreased gradually while looking for the second applanation pressure, P2.

The deformation was monitored at two points; the corneal centre and a point 0.6 mm away from the centre (the next point on the model's anterior surface away from the centre). Applanation was detected when the two points had the same coordinate in the anterior-posterior direction.

Figure 44 shows that when hysteresis was considered, the deformation under unloading differed from that under loading, as would be expected. As a result, the applanation pressures, P1 and P2, at which the central cornea flattened under air pressure, were different. However, when hysteresis was omitted, the two behaviour patterns were identical, and subsequently P2 was incorrectly equal to P1.

A similar test involving the simulation of the Goldmann Applanation Tonometer (GAT) was conducted. Unlike ORA, GAT is a contact tonometer that pushes a solid cylindrical device against the central cornea until applanation over a circular area with 3.06 mm diameter is achieved.

The behaviour was simulated numerically (including the initial application of intraocular pressure followed by contact with the tonometer), and the stress distribution during the two loading stages is depicted in Figure 45.

The figure shows how the central corneal region undergoes unloading during contact with the tonometer – a region which should follow the material's unloading behaviour in order for the simulation to provide an accurate representation of the GAT procedure.

An attempt has been made to assess the importance of considering corneal hysteresis in the simulation of GAT. Two cornea models have been developed with hysteresis considered in one and ignored in the other. Similar to ORA simulations, both models have CCT = 540 µm, PCT = 690 µm, R = 7.8 mm, p = 0.82, age = 57 years and IOP = 15 mmHg. While the first model, which considers hysteresis, estimated a GAT reading of IOP (denoted IOPG) of 15.19 mmHg, the second model which ignores hysteresis predicted an IOPG value of 13.96 mmHg. The considerable difference (8%) between the two IOPG values indicate the importance of considering hysteresis although the corneal zone expected to undergo unloading is small in this case.

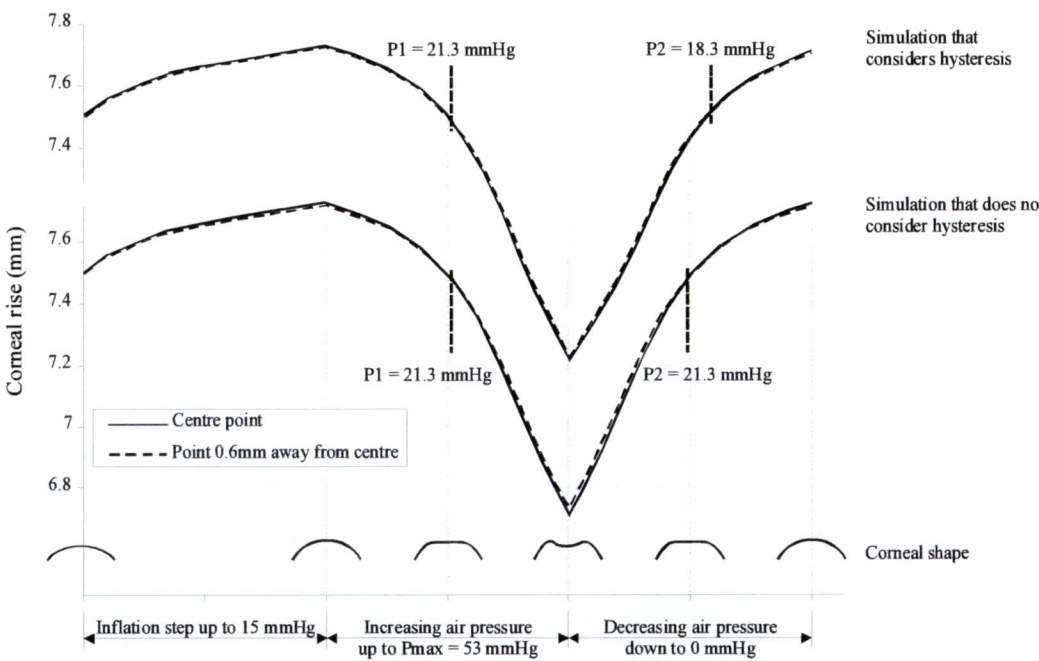

Figure 44. Corneal rise as measured at the centre and a point 0.6mm away from the centre during inflation under 15 mmHg followed by increasing then decreasing air pressure as applied by the Ocular Response Analyzer.

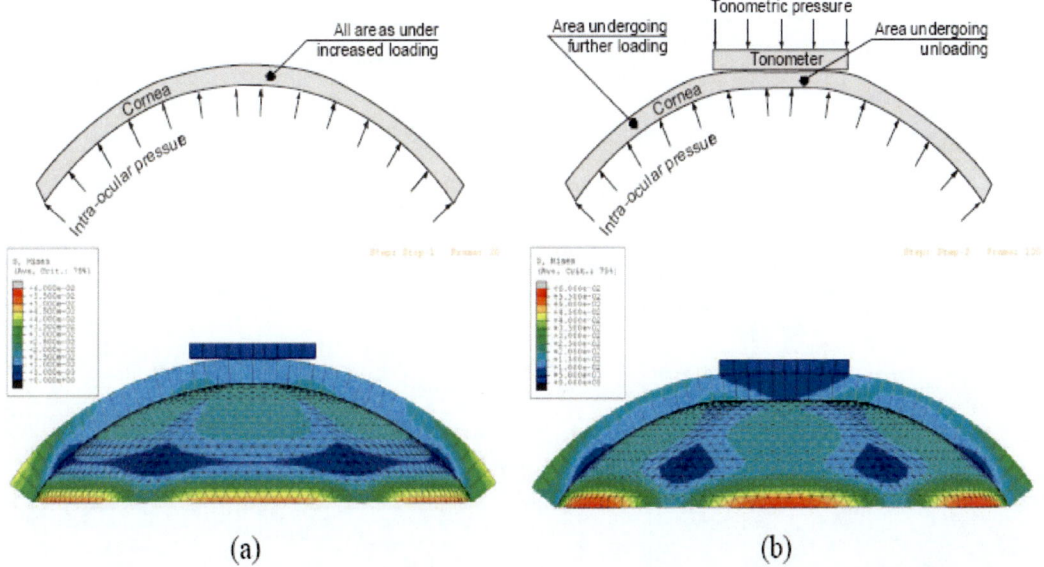

Figure 45. Stress development under IOP and GAT tonometry pressure, (a) Cornea under IOP with all areas following loading behaviour, (b) Cornea under both IOP and tonometry pressure with area under tonometer experiencing unloading.

Chapter 6

CONCLUSION

Numerical simulation has a strong potential in corneal biomechanics applications. It can be, and has been, used to improve the accuracy of tonometry [118, 119], the planning of refractive surgery [5,6, 14, 107], the evaluation of damage caused by impacts [11], the design of contact lenses, and the estimation of disease progression. This strong potential can be realised with growing interest from various research groups [120, 121], fast advances in numerical modelling, and significant contributions to knowledge and understanding of corneal microstructure, topography and material characterisation.

However, a number of challenges currently exist and slow the progress of corneal numerical simulation. First, the level of complexity of corneal microstructure with variable interweaving of stromal lamellae, and change in preferential orientation of collagen fibrils from vertical and horizontal at the centre to circumferential near the limbus, make the development of a corneal model that is truly representative of real-life conditions highly complex.

Second, the material properties of corneal tissue pose serious challenges in simulation. On a basic level, the material's hyperelasticity and hysteresis can be considered, albeit with some difficulty, in numerical simulation. However, these properties are not constant, but vary with age, rate of loading and anatomical direction, and therefore a material model which is adequate for a particular case, could be entirely inaccurate in another. One typical example is the difference in speed of loading experienced under contact and non-contact tonometry, which would require widely different material models [118, 119]. The stiffening effect of ageing adds another complication to the behaviour under both loading and unloading [30, 34].

Third and most important is the current lack of understanding of some fundamental corneal properties. These include the effect of swelling (following sleep or contact lens wear) on corneal microstructure and biomechanical behaviour, the changes in microstructure associated with the progression of diseases such as keratoconus, the effects of long term exposure to high levels of IOP in glaucoma, and the wound healing process following surgery and whether stromal tissue is able to regain its mechanical integrity after wound healing. These effects must be quantified experimentally and/or clinically before they can be accurately considered in numerical simulation models.

Fourth, the current level of understanding of other ocular components, most notably the sclera, is not adequate to enable their proper inclusion in ocular simulations. While a few studies conducted over the last 40 years helped improve understanding of scleral thickness variation, regional biomechanical performance and microstructure [122-130], work is still required to quantify the effect of sclera's hyperelasticity, hysteresis, viscoelasticity, anisotropy, layered composition, microstructure, and medical conditions such as myopia on scleral biomechanics. While this work will be important in its own right, it will enable the construction of more reliable whole eye globe models without having to approximate the boundary conditions of the cornea.

The above discussion is intended to point at the areas where corneal biomechanics research is most needed, and to draw attention to the challenges slowing the development and application of corneal numerical simulation. However, it must be emphasised that while work to address these challenges is needed, corneal numerical simulation should not wait and should progress based on current knowledge. What has been achieved over the past few decades in corneal material characterisation, topography mapping and microstructure detailing, is significant and sufficient for several important applications. Normal, healthy and intact corneas can now be reliably modelled under actions such as contact and non-contact tonometry or external impacts. These applications not only address important and relevant clinical problems, they also help demonstrate the potential value of numerical simulation in corneal biomechanics applications. Nevertheless, care should be taken not to expand simulation into areas where knowledge of material behaviour or microstructure has not yet been generated. These attempts would be likely to produce unreliable results, and might affect the general perception of what numerical simulation can do in corneal applications.

Detailed clinical validation is also required to assess the accuracy of numerical simulations and demonstrate their suitability in providing solutions to clinical problems. Clinical validation will also help overcome the traditional reluctance of ophthalmologists and optometrists to embrace predictive tools based on numerical simulations, and prepare the ground for their wide acceptance.

REFERENCES

[1] Fatt I. Physiology of the eye: An introduction to the vegetative functions, Butterworths, London, 1978.

[2] Roberts C. The cornea is not a piece of plastic. *Journal of Refractive Surgery*. 2000;16:407-413.

[3] Ethier CR, Johnson M, Ruberti J. Ocular biomechanics and biotransport. *Annual Review of Biomedical Engineering*. 2004;6:249-273.

[4] Michelacci YM. Collagens and proteoglycans of the corneal extracellular matrix. *Brazilian Journal of Medical and Biological Research*. 2003;36(8):1037-1046.

[5] Vito RP, Shin TJ, McCarey BE. A Mechanical Model of the Cornea: The Effects of Physiological and Surgical Factors on Radial Keratotomy Surgery. *Refractive Corneal Surgery*. 1989;5(2):82-88.

[6] Pinsky P, Datye V. A microstructurally-based finite element model of the incised human cornea. *Journal of Biomechanics*. 1991;10:907-922.

[7] Bryant M, McDonnell P. Constitutive laws for biomechanical modeling of refractive surgery. *Journal of Biomechanical Engineering*. 1996;118:473-481.

[8] Buzard KA. Introduction to bio-mechanics of the cornea. *Refractive and Corneal Surgery*. 1992;8:127-138.

[9] Dupps WJ. Biomechanical modeling of corneal ectasia. *J Refractive Surgery*. 2005;21(2):186-190.

[10] Velinsky S, Bryant M. On the Computer-Aided and Optimal Design of Keratorefractive Surgery. *Refractive Corneal Surgery*. 1992;8:173-183.

[11] Uchio E, Ohno S, Kudoh J, Aoki K, Kisielewicz LT. Simulation model of an eyeball based on finite element analysis on a supercomputer. *Br. J. Ophthalmology*. 1999;83:1106-1111.

[12] Fernández DC, Niazy AM, Kurtz RM, Djotyan GP, Juhasz T. Finite element analysis applied to cornea reshaping. *Journal of Biomedical Optics*. 2005;10(6):064018-1-11.

[13] Fernández DC, Niazy AM, Kurtz RM, Djotyan GP, Juhasz T. Computational modeling of corneal refractive surgery. *Proceedings of SPIE*. 2004;Vol. 5314. Bellingham, WA.

[14] Alastrué V, Calvo B, Peña E, Doblaré M. Biomechanical modeling of refractive corneal surgery. *Biomechanical Engineering*. 2006;128;150-160.

[15] Hanna, K., Jouve, F., Waring, G., and Ciarlet, P., 1992, "Computer simulation of arcuate keratotomy for astigmatism," *Refract. Corneal Surg.* 8 _1992_ 152-163.

[16] Pinsky, P., and Datye, V., 1991, "A microstructurally-based finite element model of the incised human cornea," *J. Biomech.* 10, pp. 907-922.

[17] Howland, H., Rand, R., and Lubkin, S., 1992, "A Thin-Shell Model of the Cornea and its Application to Corneal Surgery," *Refract. Corneal Surg.* 8, pp.183-186.

[18] Bryant, M. R., Machi, V., and Juhasz, T., 2000, "Mathematical Model of Picosecond Laser Keratomileusis For High Myopia," *ASME J. Biomech. Eng.* 16, pp. 155-162.

[19] Djotyan, G. P., Kurtz, R. M., Cabrera, D., and Juhasz, T., 2001, "An Analytically Solvable Model For Biomechanical Response of the Cornea to Refractive Surgery," *J. Refract. Surg.* 123, pp. 440-445.

[20] Huang, D., Tang, M., and Shekhar, R., 2003, "Mathematical Model of the Corneal Surface Smoothing After Laser Refractive Surgery," *Am. J. Ophthalmol.* 135, pp. 267-278.

[21] Yeh, H.-L., Huang, T., and Schachar, R., 2000, "A Closed Shell Structured Eyeball Model With Application to Radial Keratotomy," *J. Biomech. Eng.* 122, pp. 504–510.

[22] Hanna, K. D., Jouve, F., Bercovier, M. H., and Waring, G. O., 1989, "Computer Simulation of Lamellar Keratectomy and Laser Myopic Keratomileusis," *J. Refract. Surg.* 4, pp. 222–231.

[23] Hanna, K. D., Jouve, F., Ciarlet, P., and Waring, G. O., 1989, "Computer Simulation of Arcuate and Radial Incisions Involving the Corneoscleral Limbus," Eye 3, pp. 227–239.

[24] Boote C, Dennis S, Huang Y, Quantock AJ, Meek KM. Lamellar orientation in human cornea in relation to mechanical properties. *Journal of Structural Biology*. 2005;149:1-6.

[25] Müller LJ, Pelsa E, Schurmans LRHM, Vrensen GFJM. A new three-dimensional model of the organization of proteoglycans and collagen fibrils in the human corneal stroma. *Experimental Eye Research*. 2004;78:493-501.

[26] Meek KM, Newton RH. Organization of collagen fibrils in the corneal stroma in relation to mechanical properties and surgical practice. *Journal of Refractive Surgery*. 1999;15(6):695-699.

[27] Aghamohammadzadeh H, Newton RH, Meek KM. X-ray scattering used to map the preferred collagen orientation in the human cornea and limbus. *Structure*. 2004;12:249-256.

[28] Elsheikh A, Brown M, Alhasso D, Rama P, Campanelli M, Garway-Heath DF. Experimental assessment of corneal anisotropy. *Journal of Refractive Surgery*. 2008;24(2):178-187.

[29] Jayasuriya AC, Ghosh S, Scheinbeim JI, Lubkin V, Bennett G, Kramer P. A study of piezoelectric and mechanical anisotropies of the human Cornea. *Biosens Bioelectron*. 2003;18:381-387.

[30] Elsheikh A, Wang D, Rama P, Campanelli M, Pye D. Assessment of corneal biomechanical properties and their variation with age. *Current Eye Research*. 2007;32:11-19.

[31] Elsheikh A, Wang D, Pye D. Determination of the modulus of elasticity of the human cornea. *Journal of Refractive Surgery*. 2007;23(8):808-818.

[32] Tallec PL, Rahier C, Kaiss A. Three-dimensional incompressible viscoelasticity in large strains: Formulation and numerical approximation. *Comput. Methods Appl. Mech. Eng.* 1993;109:233-258.

[33] Boyce B, Jones R, Nguyen T, Grazier J. Stress-controlled viscoelastic tensile response of bovine cornea. *Journal of Biomechanics*. 2007; 40(11):2367-2376.

[34] Elsheikh A, Wang D, Rama P, Campanelli M, Garway-Heath DF. Experimental assessment of human corneal hysteresis. *Current Eye Research*. 2008;33:205-213.

[35] Elsheikh A, Alhasso D, Rama P. Biomechanical properties of human and porcine corneas. *Experimental Eye Research*. 2008;86:783-790.

[36] Kampmeier J, Radt B, Birngruber R, Brinkmann R. Thermal and biomechanical parameters of porcine cornea. *Cornea*. 2000;19:355-362.

[37] Scherer K, Eggert H, Guth H, Stiller P. Application of simulation techniques in human eye corneal surgery. 13th International Conference of Society for Medical Innovation and Technology. Berlin. 2001.

[38] Elsheikh A, Alhasso D, Rama P. Assessment of the epithelium's contribution to corneal biomechanics. *Experimental Eye Research*. 2008;86:445-451.

[39] Elsheikh A, Ross S, Alhasso D, Rama P. Numerical study of the effect of corneal layered structure on ocular biomechanics. *Current Eye Research*. In Press

[40] Schmack I, Dawson DG, McCarey BE, Waring BO, Grossniklaus HE, Edelhauser HF. Cohesive tensile strength of human LASIK wounds with histologic, ultrastructural and clinical correlations. *Journal of Refractive Surgery*. 2005;21:433-445.

[41] Smolek MK, McCarey BE. Interlamellar adhesive strength in human eyebank corneas. *IOVS*. 1990;31(6):1087-1095.

[42] Smolek MK. Interlamellar cohesive strength in the vertical meridian of human eye bank corneas. *IOVS*. 1993;34:2962-2969.

[43] Dubbelman M, Weeber HA, van der Heijde RGL, Völker-Dieben HJ. Radius and asphericity of the posterior corneal surface determined by corrected Scheimpflug photography. *Acta Ophthalmol. Scand.* 2002;80:379-383.

[44] Read SA, Collins MJ, Carney LG, Franklin RJ. The topography of the central and peripheral cornea. *IOVS*. 2006;47:1404-1415.

[45] Guillon M, Lydon PM, Wilson C. Corneal topography: a clinical model. *Ophthalmic. Physiol. Opt.* 1986;6:47-56.

[46] Aghaian E, Choe JE, Lin S, Stamper RL. Central Corneal Thickness of Caucasians, Chinese, Hispanics, Filipinos, African Americans, and Japanese in a Glaucoma Clinic. *Ophthalmology*. 2004;111:2211-2219.

[47] Cho P, Cheung SW. Central and peripheral corneal thickness measured with the TOPCON specular microscope SP-2000P. Current Eye Research. 2000;21(4):799-807.

[48] Lam AKC, Chan JS. Corneal thickness at different reference points from Orbscan II system. *Clinical and Experimental Optometry*. 2003;86(4):230-234.

[49] Ethier CR, JohnsonM, Ruberti J. Ocular biomechanics and biotransport. *Annu. Rev. Biomed. Eng.* 2004. 6:249-73.

[50] Pouliquen Y. 1985. Fine structure of the corneal stroma. *Cornea* 3:168-77.

[51] Parry DA, Craig AS. Electron microscope evidence for an 80 A° unit in collagen fibrils. *Nature*. 1979;282:213–215.

[52] Komai Y, Ushiki T. The three-dimensional organization of collagen fibrils in the human cornea and sclera. *Invest. Ophthalmol. Vis. Sci.* 1991;32:2244-2258.

[53] Maurice DM, Monroe F. Cohesive strength of corneal lamellae. *Exp. Eye Res.* 1990;50:59.

[54] Kokott W. Über mechanisch-funktionelle Strukturen des Auges. *Albrecht. v Grafes Arch. Ophthalmol* 1938;118:424-485.

[55] Meek KM, Blamires T, Elliot GF, Gyi TJ, Nave C. The organization of collagen fibrils in the human corneal stroma: a synchroton x-ray distraction study. *Current Eye Research,* 1987;6:841-846.

[56] Newton RH, Meek KM. The integration of the corneal and limbal fibrils in the human eye. *Biophysical Journal,* 1998;75:2508-2512.

[57] Newton RH, Meek KM. Circumcorneal annulus of collagen fibrils in the human limbus. *Investigative Ophthalmology and Visual Science.* 1998;39:1125-1134.

[58] Daxer A, Fratzl P. Collagen fibril orientation in the human corneal stroma and its implication in keratoconus. *Investigative Ophthalmology and Visual Science,* 1997;38(1):121-1293.

[59] Hayes, S., Boote, C., Lewis, J., Sheppard, J., Abahussin, M., Quantock, A.J., Purslow, C., Votruba, M., Meek, K.M., 2007. Comparative study of fibrillar collagen arrangement in the corneas of primates and other mammals. *The Anatomical Record* 290:1542-1550.

[60] Boote C, Dennis S, Meek KM. Spatial mapping of collagen fibril organisation in primate cornea—an X-ray diffraction investigation. *Journal of Structural Biology* 146 (2004) 359-367.

[61] Prangen, A., 1928. A study of the comparative anatomy of the extraocular muscles. *Trans. Am. Ophthalmol. Soc.* 26:353-380.

[62] Green K (1991) Corneal endothelial structure and function under normal and toxic conditions. *Cell Biology Review* 25, 169-207.

[63] Johnson M, Shapiro A, Ethier CR, Kamm RD. Modulation of outflow resistance by the pores of the inner wall endothelium. *IOVS.* 1992;33:1670-1675.

[64] Coster DJ. Cornea. BMJ Publishing Group, London. 2001.

[65] Patel, S., Alio, J.L., Perez-Santonja, J.J., 2004. Refractive index change in bovine and human corneal stroma before and after LASIK: A study of untreated and re-treated corneas implicating stromal hydration. *IOVS.* 45(10), 3523-3530.

[66] Beuerman RW, Pedroza L (1996) Ultrastructure of human cornea. *Microscopy Research and Technique* 33, 320-335.

[67] Danielsen CC. Tensile mechanical and creep properties of Descemet's membrane and lens capsule. *Experimental Eye Research* 79 (2004) 343-350.

[68] Merindano MD, Costa J, Canals M, Potau JM, Ruano D. A comparative study of Bowman's layer in some mammals: Relationships with other constituent corneal structures. *Eur. J. Anat.,* 6 (3) : 133-139 (2002)

[69] Seiler T, Matallana M, Sendler S, Bende T. Does Bowman's layer determine the biomechanical properties of the cornea? *Refract Corneal Surg.* 1992;8(2):139-42.

[70] Feltgen, N., Leifert, D., Funk, J., 2001, "Correlation between central corneal thickness, applanation tonometry and direct intracameral IOP readings," *British Journal of Ophthalmology,* 85, pp. 85-87.

[71] Laiquzzaman, M., Bhojwani, R., Cunliffe, I., Shah, S., 2006, "Diurnal variation of ocular hysteresis in normal subjects: relevance in clinical context," *Clinical and Experimental Ophthalmology,* 34, pp. 114-118.

[72] Foster, P. J., Machin, D., Wong, T.-Y., Ng, T.-P., Kirwan, J. F., Johnson, G. J., Khaw, P. T., Seah, S. K. L., 2003, "Determinants of intraocular pressure and its association with glaucomatous optic neuropathy in Chinese Singaporeans: The Tanjong Pagar study," *IOVS,* 44, pp. 3885-3891.

[73] Lleó A, Marcos A, Calatayud M, Alonso L, Rahhal SM, Sanchis-Gimeno JA. The relationship between central corneal thickness and Goldmann applanation tonometry. *Clin. Exp. Optom.* 2003;86(2):104-108.

[74] Shah, S., Spedding, C., Bhojwani, R., Kwartz, J., Henson, D., McLeod, D., 2000, "Assessment of the diurnal variation in central corneal thickness and intraocular pressure for patients with suspected glaucoma," *Ophthalmology*, 107(6), pp. 1191-1193.

[75] Hirji, N. K., Larke, J. R., 1978, "Thickness of human cornea measured by topographic tachometry," *American Journal of Optometry*, 55, pp. 97-100.

[76] Martola, E. L., Baum, J. L., 1968, "Central and peripheral corneal thickness," *Arch. Opthhalmol*, 79, pp. 28-30.

[77] Maurice, D. M., Giardini, A. A., 1951, "A simple optical apparatus for measuring the corneal thickness and the average thickness of the human cornea," *British Journal of Ophthalmology,* 35, pp. 169-177.

[78] Hennighausen H, Feldman ST, Bille JF, McCulloch AD. Anterior-posterior strain variation in normally hydrated and swollen rabbit cornea. IOVS. 1998;39:253-262.

[79] Nash IS, Greene PR, Foster CS. Comparison of mechanical properties of keratoconus and normal corneas. *Exp. Eye Res.* 1982; 35:413-24.

[80] Fung, Y.C. Biomechanics: Mechanical properties of Living Tissues. Springer-Verlag, New York, pp. 211, 1981.

[81] Hjortdal, J.O. "Regional elastic performance of the human cornea." *J. Biomechanics*, Vol. 29, No. 7, pp. 931-942, 1996.

[82] Shin, T.J., Vito, R.P. et al. "The distribution of strain in the human cornea." *J. Biomechanics,* Vol. 30, No. 5, pp. 497-503, 1997.

[83] Erkamp R.Q., Wiggins P, Skovoroda, A.R. et al. "Measuring the elastic modulus of small tissue samples." *Ultrasound Imaging*, 20:17–28, 1998.

[84] Chen E.J., Novakofski J., Jenkins W.K., et al. "Young's modulus measurements of soft tissues with application to elasticity imaging." *IEEE Trans. UFFC* 43:191–4, 1996.

[85] Zeng, Y., Yang, J., et al. "A comparison of biomechanical properties between human and porcine cornea." *J. Biomechanics*, Vol. 34, No. 5, pp. 533-537, 2001.

[86] Jue B, Maurice DM. The mechanical properties of rabbit and human cornea. *J. Biomech.* 1986;19:847–853.

[87] Woo SLY, Kobayashi AS, Schlegel WA, Lawrence C. Non-linear Properties of intact cornea and sclera. *Exp. Eye Res.* 1972; 14: 29-39.

[88] Nyquist GW. Rheology of the cornea: Experimental techniques and results. *Experimental Eye Research*. 1968; 7: 183-188.

[89] Hjortdal JO. On the biomechanical properties of the cornea with particular reference to refractive surgery. *The Ophthalmological Journal of the Nordic Countries*. 1998; 76(225): 1-23.

[90] Andreassen TT, Simonsen AH, Oxlund H. Biomechanical properties of keratoconus and normal corneas. *Exp. Eye Research*. 1980; 31: 435-441.

[91] Liu, J., Roberts, C. J., 2005, "Influence of corneal biomechanical properties on intraocular pressure measurement: Quantitative analysis," *J. Cataract Refract Surgery*, 31, pp. 146-155.

[92] Hoeltzel DA, Altman P, Buzard K, Choe K. Strip extensiometry for comparison of the mechanical response of bovine, rabbit, and human corneas. *J. Biomech. Eng.* 1992;114(2):202–215.

[93] Daxer A, Misof K, Grabnerj B, Ettl A, Fratzl P. Collagen fibrils in the human corneal stroma: Structure and aging. *Investigative Ophthalmology and Visual Science*, 1998;39(3):644-648.

[94] Malik NS, Moss SJ, Ahmed N, Furth AJ, Wall RS, Meek KM, Ageing of the human corneal stroma: structural and biochemical changes, *Biochimica et Biophysica Acta.* 1992;1138:222-228.

[95] Wollensak G, Spoerl E, Seiler T. Stress-strain measurements of human and porcine corneas after riboflavin–ultraviolet-A-induced cross-linking. *Journal of Cataract and Refractive Surgery.* 2003;29:1780-1785.

[96] Anderson, K., Elsheikh, A. and Newson, T., Application of Structural Analysis to the Mechanical Behaviour of the Cornea, *Journal of Royal Society – Interface*, 1, 2004, 1-13

[97] Greene, P.R. "Stress-strain behavior for curved exponential strips." *Bulletin of Mathematical Biology*, Vol. 47,No. 6, pp. 757-764, 1985.

[98] Elsheikh A, Anderson K. Comparative study of corneal strip extensometry and inflation tests. *Journal of Royal Society – Interface.* 2005;2:177–185.

[99] Naor J, Slomovic AR, Chipman M, Rootman DS. A randomized, double-masked clinical trials of Optisol-GS vs Chen medium for human corneal storage. *Arch. Ophthalmology.* 2002;120:1280-1285.

[100] Arndt C, Reese S, Kostlin R. Preservation of canine and feline corneoscleral tissue in Optisol GS. *Veterinary Ophthalmology.* 2001;4(3):175-182.

[101] Kaufman HE, Beuerman RW, Steinemann TL, Thompson HW, Varnell ED. Optisol corneal storage medium. *Arch. Ophthalmol.* 1991;109(6):864-868.

[102] Wagoner MD, Gonnah E. Corneal graft survival after prolonged storage in Optisol-GS. *Cornea.* 2005;24(8): 976-979.

[103] Orssengo GJ, Pye DC. Determination of the true intraocular pressure and modulus of elasticity of the human cornea in vivo. *Bull Math. Biol.* 1999; 61:551-572.

[104] Elsheikh, A. and Alhasso, D., Mechanical Anisotropy of Porcine Cornea and Correlation with Stromal Microstructure, Accepted for publication in Experimental Eye Research.

[105] Bryant MR, Szerenyi K, Schmotzer H, McDonnell P. Corneal tensile strength in fully healed radial keratotomy wounds. *IOVS.* 1994;35(7):3022-3031.

[106] Elsheikh A, Wang D. Numerical modelling of corneal biomechanical behaviour. *Computer Methods in Biomechanics and Biomedical Engineering.* 2007;10(2):85-95.

[107] Fernández DC, Niazy AM, Kurtz RM, Djotyan GP, Juhasz T. A finite element model for ultrafast laser-lamellar keratoplasty. *Annals of Biomedical Engineering.* 2006;34(1):169-183.

[108] Radner W, Zehetmayer M, Skorpik C, Mallinger R. Altered organization of collagen in the apex of keratoconus corneas. *Ophthalmic Research.* 1998;30(5):327-332.

[109] Meek KM, Tuft SJ, Huang Y, Gill PS, Hayes S, Newton RH, Bron AJ. Changes in collagen orientation and distribution in keratoconus corneas. *IOVS.* 2005;46:1948-1956.

[110] Luce DA. Determining in vivo biomechanical properties of the cornea with an ocular response analyzer. *Journal of Cataract Refractive Surgery.* 2005;31:156-162.

[111] Shah S, Laiquzzaman M, Bhojwani R, Mantry S, Cunliffe I. Assessment of the biomechanical properties of the cornea with the Ocular Response Analyzer in normal and keratoconic eyes. *IOVS.* 2007;48:3026-3031.

[112] Zienkiewicz OC, Taylor RL. The finite element method. McGraw-Hill Book Company Europe, Berkshire. 2000.

[113] Hamri O, Léon J-C, Giannini F, Falcidieno B. Exploiting features for finite element model generation. *Mathematics in Industry.* 2006;8:585-589.

[114] Armstrong CG. Modelling requirements for finite-element analysis. *Computer-aided design.* 1994;26(7):573-578.

[115] Babuška I, Strouboulis T. The finite element method and its reliability. Oxford Science Publications. 2001.

[116] McPhee TJ, Bourne WM, Brubaker RF. Location of the stress-bearing layers of the cornea. *IOVS.* 1985;26:869-872.

[117] Kotecha A, Elsheikh A, Roberts C, Zhu H, Garway-Heath DF. Corneal thickness- and age-related biomechanical properties of the cornea measured with the Ocular Response Analyzer. *IOVS.* 2006;47(12):5337-5347.

[118] Elsheikh, A., Alhasso, D., Kotecha, A. and Garway-Heath, D.F., Assessment of the Ocular Response Analyzer as a tool for intraocular pressure measurement, *Journal of Biomechanical Engineering.* In Press.

[119] Elsheikh, A., Wang, D., Kotecha, A., Brown, M. and Garway-Heath, D., Evaluation of Goldmann applanation tonometry using a nonlinear finite element ocular model, *Annals of Biomedical Engineering*, 34(10), 2006, 1628-1640.

[120] Pandolfi A, Manganiello F. A model for the human cornea: constitutive formulation and numerical analysis. *Biomechanics and modeling in mechanobiology.* 2006;.5(4):237-246.

[121] Pinsky PM, van der Heide D, Chernyak D. Computational modeling of mechanical anisotropy in the cornea and sclera. *Journal of Cataract and Refractive Surgery.* 2005;31:136-145.

[122] Curtin BJ, Physiopathologic aspects of scleral stress-strain. *Trans. Am. Ophthalmol. Soc.* 1969;67:417-461.

[123] Philips JR, McBrien NA. Form deprivation myopia: elastic properties of sclera. *Ophthal. Physiol. Opt.* 1995;15(5):357-362.

[124] Downs JC, Suh JKF, Thomas KA, Bellezza AJ, Hart RT, Burgoyne CF. Viscoelastic material properties of the peripapillary sclera in normal and early-glaucoma monkey eyes. *IOVS.* 2005;46:540-546.

[125] Downs JC, Suh JKF, Thomas KA, Bellezza AJ, Burgoyne CF, Hart RT. Viscoelastic characterization of peripapillary sclera: material properties by quadrant in rabbit and monkey eyes. *Journal of Biomechanical Engineering.* 2003;125:124-131.

[126] Barraglioli JL, Kamm RD. Measurements of the compressive properties of sclerol tissue. *IOVS.* 1984;25:59-65.

[127] Spoerl E, Boehm AG, Pillunat LE. The influence of various substances on the biomechanical behavior of lamina cribrosa and peripapillary sclera. *IOVS.* 2005;46:1286-1290.

[128] Norton TT, Rada JA. Reduced extracellular matrix in mammalian sclera with induced myopia. *Vision Research.* 1995;35(9):1271-1281.

[129] Guggenheim JA, McBrien NA. Form-deprivation myopia induces activation of scleral matrix metalloproteinase-2 in tree shrew. *IOVS*. 1996;37:1380-1395.

[130] McBrien NA, Gentle A. Role of the sclera in the development and pathological complications of myopia. *Progress in Retinal and Eye Research*. 2003;22(3):307-338.

INDEX

A

accuracy, ix, 1, 18, 35, 37, 38, 51, 52
adhesion, 2, 44
adhesive strength, 55
African Americans, 55
age, 11, 20, 21, 22, 23, 24, 25, 27, 31, 35, 47, 48, 51, 54, 59
ageing, 11, 51, 58
anatomy, 56
anisotropy, 2, 9, 11, 14, 20, 26, 27, 45, 52, 54, 58, 59
aqueous humor, 4
assessment, 2, 27, 30, 46, 54, 55
astigmatism, 53
attachment, 5

B

behavior, 58, 59
bending, 38, 40, 43, 44
biomechanics, ix, 1, 2, 13, 20, 27, 35, 43, 51, 52, 53, 55
boundary conditions, 13, 37, 38, 52
bovine, 54, 56, 58
buttons, 11, 17, 19

C

Caucasians, 55
clinical trials, 58
clinical validation, 52
collagen, 2, 3, 4, 5, 9, 17, 26, 45, 51, 54, 55, 56, 58
combined effect, 15
complexity, 38, 51
complications, 60
components, 1, 2, 9, 11, 18, 51
computer simulation, 53
costs, 38
creep, 2, 9, 20, 30, 31, 32, 56
creep tests, 30

cross-linking, 9, 11, 58
cross-sectional, 13

D

deformation, 4, 5, 9, 11, 13, 17, 18, 23, 28, 30, 37, 39, 40, 41, 44, 48
degradation, 17
deprivation, 59, 60
deviation, 12, 42
digital cameras, 18
direct measure, 25
disease progression, 51
diseases, 51
distraction, 56
distribution, 2, 4, 13, 14, 15, 16, 37, 43, 47, 48, 57, 58
donor, 11, 20, 23, 25, 27, 35
drying, 17
duration, 17

E

elasticity, 54, 57, 58
elongation, 13, 14, 15, 34
endothelium, 3, 4, 5, 9, 35, 37, 38, 43, 56
epithelium, 2, 3, 5, 9, 20, 27, 28, 29, 35, 37, 38, 43, 55
equilibrium, 37
exclusion, 29
execution, 38
exercise, 38, 41
exposure, 51
extracellular matrix, 3, 4, 53, 59
extraction, 19
extraocular muscles, 4, 56
eyeball, 53
eyes, 5, 35, 59

F

fibrillar, 4, 56
fibrils, 2, 3, 4, 9, 13, 17, 26, 45, 51, 54, 55, 56, 58
film, 5, 27
finite element method, 59
flow, 17
fluid, 4, 5
fluid transport, 4

G

glaucoma, 51, 56, 57, 59
glycoproteins, 3

H

healing, 51
heterogeneous, 43
Hispanics, 55
homogenous, 1, 29, 43
hydration, 11, 17, 19, 56
hysteresis, 2, 20, 25, 35, 46, 48, 51, 55, 56

I

images, 4
imaging, 57
in vivo, 12, 35, 58
inclusion, 51
incompressible, 12, 30, 54
indication, 7
inflation, 11, 12, 13, 15, 17, 18, 19, 20, 22, 31, 49, 58
injury, ix, 1, 5, 35
interaction, 44
interface, 29
intraocular pressure (IOP), 13, 17, 20, 22, 38, 47, 48, 56, 57, 58, 59
isotropic, 45
isotropy, 26, 30

K

keratinocytes, 3
keratoplasty, 58

L

lamella, 4, 9, 35
lamellae, 3, 4, 5, 9, 28, 29, 30, 44, 51, 55
lamellar, 2, 29, 44, 58
lamina, 44, 59
laminar, 44
laser, 18, 58
leakage, 4

lenses, ix, 9, 35, 51, 56
limbus, 4, 8, 9, 11, 12, 40, 51, 54, 56
limitations, 14
linear, 1, 16, 22, 23, 38, 42, 57
localised, 38, 40

M

mapping, 2, 35, 52, 56
matrix, 3, 4, 37, 53, 59, 60
matrix metalloproteinase, 60
measurement, 35, 57, 59
measures, 18
mechanical properties, 4, 9, 54, 57
median, 12, 15, 40
meridian, 55
microscope, 55
microstructure, 2, 5, 9, 11, 26, 35, 51, 52
modeling, 53, 59
models, 1, 2, 10, 20, 35, 37, 38, 39, 40, 41, 42, 45, 46, 48, 51, 52
modulus, 10, 12, 22, 23, 30, 42, 44, 54, 57, 58
monolayer, 4
Monroe, 55
movement, 29, 41
muscles, 4
myopia, 35, 52, 59, 60

N

neuropathy, 56
nodes, 37
non-uniform, 13, 20, 37
normal, 7, 11, 12, 20, 35, 56, 57, 59
numerical analysis, 59
nutrients, 4, 5, 27

O

ophthalmologists, 52
optical, 1, 57
orientation, 2, 4, 9, 13, 19, 26, 45, 51, 54, 56, 58
oxygen, 5, 27

P

parameter, 35, 38
patients, 57
PCT, 8, 12, 47, 48
perception, 52
personal computers, 2
physiological, 5
piezoelectric, 54
Poisson, 12, 30
polynomial, 21

prediction, 37
predictive model, ix
pressure, 11, 12, 13, 17, 18, 19, 20, 21, 22, 23, 24, 25, 26, 28, 29, 31, 38, 39, 40, 41, 42, 43, 44, 45, 46, 47, 48, 49, 56, 57, 58, 59
proteoglycans, 3, 53, 54
pulse, 48

R

radial keratotomy, 58
radius, 7, 12, 15, 37, 40
range, 8, 10, 11, 17, 20, 23, 34
reading, 48
reality, ix, 38
refractive index, 1
relaxation, 2, 20, 30, 31, 32, 33
relevance, 56
reliability, 13, 38, 59
remodelling, 9
reservoir, 17, 18
resistance, 4, 5, 9, 39, 44, 56
resolution, 18
riboflavin, 58
rings, 37, 38

S

saline, 17
scattering, 54
sclera, 12, 35, 37, 41, 51, 55, 57, 59, 60
sensitivity, 42
separation, 32
shape, 1, 7, 37, 40
shear, 2, 29, 30, 44
Shell, 54
simulation, ix, 2, 25, 29, 33, 38, 41, 46, 48, 51, 52, 53, 55
sleep, 51
stages, 32, 48
standard deviation, 7, 8
statistical analysis, 29, 31, 32
stiffness, 4, 5, 9, 11, 20, 21, 22, 24, 25, 26, 28, 29, 34, 39, 40, 42, 43, 44, 45
storage, 58
strain, 2, 10, 11, 12, 13, 14, 15, 16, 17, 20, 21, 22, 23, 24, 25, 26, 30, 31, 32, 33, 34, 42, 44, 57, 58, 59
strength, 29, 30, 55, 58
stress, 2, 10, 11, 12, 13, 15, 16, 17, 19, 20, 21, 22, 23, 24, 25, 26, 30, 31, 32, 33, 37, 40, 42, 44, 46, 47, 48, 59
stress level, 42, 46

stress-strain curves, 25
stroma, 2, 3, 4, 5, 9, 26, 27, 28, 29, 35, 37, 38, 43, 45, 54, 55, 56, 58
stromal, 2, 3, 4, 5, 9, 11, 20, 26, 28, 29, 30, 44, 45, 51, 56
substances, 59
surface tension, 30
surgeries, ix, 2
surgery, ix, 1, 2, 5, 29, 35, 51, 53, 55, 57, 58, 59
surgical intervention, 35
swelling, 4, 5, 9, 11, 27, 51

T

TEM, 4
temperature, 11, 17, 18, 19
temporal, 4, 45
tensile strength, 29, 30, 55, 58
tension, 2, 13, 15, 29, 40, 43, 44
three-dimensional model, 54, 55
tissue, 9, 10, 11, 12, 13, 14, 17, 19, 20, 24, 25, 29, 30, 33, 35, 37, 42, 46, 47, 51, 57, 58, 59
tonometry, ix, 1, 35, 49, 51, 52, 56, 57, 59
topographic, ix, 57
toxic, 56
transducer, 17
transition, 9

U

ultraviolet, 58
uniaxial tension, 19, 33, 34
uniform, 11, 13, 17, 19, 20, 37, 38, 39, 40, 41, 42, 43, 44, 45

V

validation, 38, 52
values, 7, 8, 16, 21, 22, 24, 28, 29, 32, 33, 40, 43, 45, 47, 48
variation, 4, 9, 10, 13, 14, 15, 16, 17, 26, 39, 40, 51, 54, 56, 57
viscoelastic properties, 20

W

wear, 9, 51
working conditions, 12
wound healing, 51

X

X-ray diffraction, 56